Chemistry and Science Fiction

Karsten Müller

Chemistry and Science Fiction

What We Can Learn from the Future

Karsten Müller
Institute of Technical Thermodynamics
University of Rostock
Rostock, Germany

ISBN 978-3-662-70378-6   ISBN 978-3-662-70379-3 (eBook)
https://doi.org/10.1007/978-3-662-70379-3

Translation from the German language edition: "Chemie und Science Fiction" by Karsten Müller, © Der/die Herausgeber bzw. der/die Autor(en), exklusiv lizenziert an Springer-Verlag GmbH, DE, ein Teil von Springer Nature 2022. Published by Springer Berlin Heidelberg. All Rights Reserved.

This book is a translation of the original German edition "Chemie und Science Fiction" by Karsten Müller, published by Springer-Verlag GmbH, DE in 2022. The translation was done with the help of an artificial intelligence machine translation tool. A subsequent human revision was done primarily in terms of content, so that the book will read stylistically differently from a conventional translation. Springer Nature works continuously to further the development of tools for the production of books and on the related technologies to support the authors.

© The Editor(s) (if applicable) and The Author(s), under exclusive license to Springer-Verlag GmbH, DE, part of Springer Nature 2025

TM & © 2025 CBS Studios Inc. STAR TREK and all related marks and logos are trademarks of CBS Studios Inc. All Rights Reserved.

This work is subject to copyright. All rights are solely and exclusively licensed by the Publisher, whether the whole or part of the material is concerned, specifically the rights of translation, reprinting, reuse of illustrations, recitation, broadcasting, reproduction on microfilms or in any other physical way, and transmission or information storage and retrieval, electronic adaptation, computer software, or by similar or dissimilar methodology now known or hereafter developed.
The use of general descriptive names, registered names, trademarks, service marks, etc. in this publication does not imply, even in the absence of a specific statement, that such names are exempt from the relevant protective laws and regulations and therefore free for general use.
The publisher, the authors and the editors are safe to assume that the advice and information in this book are believed to be true and accurate at the date of publication. Neither the publisher nor the authors or the editors give a warranty, expressed or implied, with respect to the material contained herein or for any errors or omissions that may have been made. The publisher remains neutral with regard to jurisdictional claims in published maps and institutional affiliations.

This Springer imprint is published by the registered company Springer-Verlag GmbH, DE, part of Springer Nature.
The registered company address is: Heidelberger Platz 3, 14197 Berlin, Germany

If disposing of this product, please recycle the paper.

Preface

Science fiction is not just exciting entertainment. Good science fiction offers not only fiction but also science. A particularly good example of this are the series and films of the Star Trek universe. For more than half a century, Star Trek has fascinated people. In the encounter with extraterrestrial cultures, we reflect on our society. The stories from distant worlds encourage a new perspective on our own world. Questions of human life, which we would otherwise hardly think about, come into focus. Since the mid-1960s, Star Trek has repeatedly accompanied and even advanced social developments. The diverse series and films, which continue to emerge to this day, are a reflection of our time and at the same time a guide to how things can continue. Therefore, Star Trek may be the greatest science fiction series that has ever existed.

It all began almost 60 years ago when the original series, known in German as *Raumschiff Enterprise*, started (in English often called: *The Original Series,* from which the abbreviation TOS, used not only in this book, originates). Captain Kirk, Mr. Spock, and Dr. McCoy traveled with the crew of the Enterprise to distant planets and experienced fantastic adventures. Then there was a gap. After the short-lived attempt of a first Star Trek animated series, there were no new adventures for a decade and a half. At least not in series form. About ten years after the end of the original series, Captain Kirk returned. This time not on television, but on the big screen. A whole series of movies followed before things finally picked up again at the end of the 1980s. Captain Picard stood on the bridge of a new Enterprise. Set a century later, in the 24th century, this Enterprise once again traveled to worlds that no human had seen before. Before *Star Trek: The Next Generation* (TNG) ended, the third live-action series followed in the early 1990s with *Star Trek: Deep Space Nine* (DS9). A series with a daring new concept. Instead of being set on a spaceship, the action took place on a space station. What initially sounds like a marginal difference is a real challenge narratively for a science fiction series. Because now the protagonists could no longer simply fly to the aliens themselves. They had to come to them. The concept worked surprisingly well, and the Star Trek universe was enriched by a whole new kind of stories. The next series, *Star Trek: Voyager* (VOY), started just two years later and was again based on a spaceship. However, this one had been stranded at the other end of the galaxy, and the journey home took a full seven years. When Voyager returned to Earth at the beginning of the 21st century (of course, it actually returned in the second half of the 24th century, but it was seen on television at the beginning of the 21st century), the fifth series went back two centuries. *Star Trek: Enterprise* (ENT) depicted the time immediately before the founding of the United Federation of Planets. When their journey across the screens ended after only four years, it almost seemed as if the story of Star Trek had also come to an end. There were still some movies, but the era of the series was over. Again, there was a dark time that lasted over a decade. But in 2017, the sixth series, *Star Trek: Discovery* (DSC), finally started, and not long after, Captain Picard also returned—in a series named after him (*Star Trek: Picard*, PIC). So the journey continues.

Unlike in some other well-known science fiction series, „science" always played a significant role in Star Trek (not just *space fantasy*). What most people think of in this context is physics. There are many exciting questions in this regard: Can a warp drive really exist? And what does Einstein say about it? How can beaming work? And what does Heisenberg say about it? Can one escape from the event horizon of a singularity? And what does Schwarzschild say about it? A whole series of books has already been written about Star Trek and physics. The technology of Star Trek is also more than exciting. Captain Kirk liked to stand on the bridge. But in a certain way, Scotty was the secret hero. Engineers have always played an essential role in Star Trek.

Admittedly, Star Trek did not make me become an engineer. That would probably have happened anyway. Only without Star Trek, I would have likely studied electrical engineering instead of chemical engineering. How does Star Trek make one specialize in chemistry?

In the years before I graduated from school, the fourth live-action series was airing—*Star Trek: Voyager*. Voyager was probably the most in love with futuristic technology of all the Star Trek series. One detail of Voyager's technology ultimately led me to choose chemistry: its computer. The computer systems of Voyager were not based on conventional computerchips but on so-called bioneural gel packs. Biochemical components formed the heart of the most technologically advanced spaceship in the Star Trek universe. That was the decisive impetus: the realization that chemistry is not just a marginal topic dealing with things like paints and lubricants. Chemistry is the key. And Star Trek can open up a whole new perspective on chemistry.

After having dealt with it scientifically for many years, I can say: Star Trek was right. Chemistry is the key to an incredible number of things. To energy technology, to medicine, to modern electrical engineering, to the understanding of nature and climate, and much more. With this book, I want to try to pass on some of that. The chemistry, as we experience it in Star Trek, is meant to serve as a bridge to approach selected chemical questions in the following chapters and hopefully gain one or two new and interesting insights.

Prof. Dr. Karsten Müller

Acknowledgments

Various people have inspired me to write this book and have contributed in different ways to its creation. Starting with Magdalena Mikulaschek, who gave me *Die Star Trek Physik* by Metin Tolan and thus first brought me the idea for this book, to my editor Désirée Claus, who accompanied the publication process, many people have made valuable contributions at the intermediate stages.

Not to be forgotten at this point are the diligent contributors who help create the wiki on the internet platform Memory Alpha. While writing, this database was very helpful as a memory aid. As a Wikipedian, I appreciate this work.

I would especially like to thank all those who contributed to clearing this book of many errors and misunderstandings by diligently proofreading various chapters. In this context, Dr. Patrick Adametz, Alexander Fendt, Dr. Christoph Krieger, Christof Müller, Dr. Peter Schulz, Dr. Susanne Spörler, and Raphael Wittenburg should be mentioned. Any remaining errors are solely my responsibility.

Contents

1. **The Chemistry of Extraterrestrial Life Forms** 1
   1.1 Horta or Life from Silicon 1
   1.2 Very Hot Extraterrestrials 8
   1.3 Life without a Body 16
   1.4 Crossing the Threshold 19
2. **Hydrogen and the Infinite Vastness** 25
   2.1 Breathing Hydrogen 25
   2.2 The Bussard Collector or Collecting from the Vacuum ... 33
   2.3 Explosions in Space 41
   2.4 One Moon Circles 48
3. **Atoms in a Completely Different Way** 55
   3.1 When Atoms Burn 55
   3.2 Tiny Atoms 62
   3.3 Tiny Atoms—Part 2 69
4. **Chemistry and Its Speed** 77
   4.1 The Salt Vampire of M-113 77
   4.2 A Thirsty Virus 85
   4.3 Simply Being Someone Else 93
   4.4 Why Does a Dead Shapeshifter Revert to Its Natural Form? 99
5. **New Materials in the 23rd and 24th Century** 107
   5.1 How many Elements are there Actually? 107
   5.2 Materials That Do Not Consist of Chemical Elements 114
   5.3 The Mixing Ratio of Matter and Antimatter 123
6. **Particularly Impressive Chemicals** 127
   6.1 Corbomite or Kirk's Favorite Chemical Bluff 127
   6.2 The Molecule of Molecules 133

Bibliography 141

The Chemistry of Extraterrestrial Life Forms

1.1 Horta or Life from Silicon

In 1967 and 1967, the first Star Trek series (often referred to as "The Original Series," TOS) began airing. This included the very well-known 26th episode of the 1st season, *"The Devil in the Dark"*. It is about how the starship Enterprise travels to the planet Janus VI in the year 2267. In the mining colony there, deaths occur repeatedly. The colonists are attacked and killed by an unknown life form. As is often the case with colonizers, there is a fundamental lack of understanding of the peculiarities of the indigenous population in this situation as well. The spheres found everywhere in the caves are not geological curiosities that can be treated completely carelessly. They are the eggs of an intelligent, indigenous life form called Horta. Naturally, Horta is not thrilled that the miners are killing their offspring (albeit unknowingly) or placing them as decorations on shelves. Accordingly, Horta defends itself, leading to the aforementioned deaths among the colony's inhabitants. The scientifically special aspect of this event is that Horta is the first case of a silicon-based life form known to the Federation.

Silicon-based life forms are a popular subject both in the context of scientific speculation and in various examples of science fiction literature. Life, as we know it from Earth, is known to be based on the element carbon. Chemically speaking, more precisely: on functionalized hydrocarbons. What does that mean? To understand life based on silicon, we first need to look at life based on carbon.

First of all, terrestrial living beings do not primarily consist of carbon. Most living beings are composed of about nine-tenths water, a compound of the elements hydrogen and oxygen. Water is essential for life—not just on Earth, but essentially everywhere. For the actual biochemistry that ensures we are not just bags filled with water but real, complex living beings, carbon compounds are ultimately needed. Water is still essential for every living being. This has a very simple reason: it is liquid. This may sound trivial, but it is of central importance. The more complex compounds based on a carbon framework, from which living beings are built, are almost all solids. They can be dissolved in water and thus handled like liquids. In their pure form, however, they are solid. If living beings did not consist mostly of water but only of these solids, no movement would be possible. This would not only make higher life forms like most animals impossible. Even immobile life forms like most plants rely on movement. There is not only the external movement that animals use to move towards their food or to avoid becoming food for other animals.

First of all (and this is the most important thing), there are internal movement processes. Without these, life would not be conceivable at all. All kinds of substances must move from one place to another inside the cell. Molecules that serve as food and are absorbed from the outside must be transported inside. Conversely, waste products must be transported out of the cell. Proteins formed inside at the so-called ribosomes must reach where they are needed. For reproduction, a cell must divide. To do this, the cell membrane must deform, constrict, and close again into two independent cells. All these processes require that molecules and at least smaller particles can move. If a cell were not filled with a liquid, this would simply not be possible. If it were instead filled with a gas, the gas (unlike a solid) would not be an obstacle to substance transport. On the other hand, it would not support the transport either, which liquids can certainly do. Many substances are soluble in water. And even non-water-soluble particles can at least float in a liquid, provided their density is roughly the same as that of water. Due to the large density difference, this is hardly possible with a gas. Therefore, all living beings ultimately depend on being filled with a liquid.

Theoretically, one could of course think that any other liquid could just as well be suitable for this. This is conceivable. However, water offers, besides the fact that it is liquid at the temperatures that mostly prevail on Earth and its great availability, other advantages. Its special chemical properties make it excellently suited to support life. For one, the water molecule is very polar. This means that the molecule is electrically neutral. However, there is an internal charge distribution within the molecule. The water molecule has a positively charged side (at the two hydrogen atoms) and a negatively charged side (at the oxygen atom). This makes it very good at dissolving substances like sugar and other polar molecules. Additionally, it is a so-called ampholyte. This means that it can both accept and donate hydrogen ions (also called protons). This allows it to support many important chemical reactions. Overall, water has a whole range of special, sometimes quite unusual properties. If the laws of nature were not exactly such that water possessed the special combination of properties it has, it would be fatal for life.

So much for water. What about carbon? Why do we even say that terrestrial life is based on carbon? After all, it is about hydrocarbons, which also contain many oxygen atoms (and to a lesser extent also nitrogen and sulfur). So why do we always say that life is based on carbon (and not water)?

There is a simple reason for this. Carbon is one of the few tetravalent elements. A carbon atom can be chemically bonded to up to four other atoms simultaneously. Hydrogen, on the other hand, is monovalent. Each hydrogen atom can only be bonded to a single other atom. Accordingly, it would be completely impossible to build more complex compounds based on hydrogen. Hydrogen atoms can certainly be part of complex molecules. But wherever there is a hydrogen atom, the structure does not go any further. If the hydrogen atom is bonded to a molecule, then its one bond is already used up. It cannot form another true chemical bond without separating from the main molecule.

> **A few more details**
>
> Hydrogen atoms that are, for example, bonded to oxygen atoms can indeed form a kind of second bond in a certain way. This is referred to as the so-called hydrogen bond. Hydrogen bonds arise because a positive charge is concentrated on the hydrogen atom. This partial positive charge is attracted by negative charges, for example, on oxygen atoms in other molecules. These hydrogen bonds represent a very strong interaction between molecules. However, they are still not strong enough to bind the two molecules so strongly that they would become one large molecule. Therefore, the statement that hydrogen is monovalent is quite justified. ◄

Hydrogen (or other monovalent elements like chlorine or bromine) can therefore play an important role in complex molecules. However, it cannot form the backbone of complex molecules. For this, an element is needed that can form multiple bonds simultaneously. Oxygen, for example, is divalent and is thus able to be bonded to up to two other atoms simultaneously. Therefore, an oxygen atom does not necessarily represent the end of a larger molecule. The chain of atoms can indeed continue after an oxygen atom. However, that is about it. Divalent atoms could only bind long chains.[1] More complex molecules, which are needed to enable life, however, require branches or at least the possibility to attach something else to the chain (a so-called *functional group*). A long series consisting only of the same atoms offers little room to form complex biomolecules. For this, the corresponding atom must be able to form at least a third bond.

An example of a trivalent element would be nitrogen. However, its chemical properties are ultimately not really suitable for forming the backbone of complex biochemical molecules. On the one hand, a long chain of nitrogen atoms would be unstable. A molecule based on a long chain of nitrogen atoms would quickly disintegrate. On the other hand, nitrogen tends to exhibit basic behavior in chemical compounds[2]. Ammonia is a well-known example of a basic nitrogen-based

---

[1] In the case of oxygen, even that would be only possible to a limited extent. If one bonds oxygen atom to oxygen atom, you get a peroxide. These peroxides are chronically unstable and even tend to explode.

[2] Basic (or often called alkaline) is the counterpart of acidic in chemistry. There are various definitions of bases and acids. The probably best-known definition is that according to Brønsted. A Brønsted base is a substance that can accept positively charged hydrogen ions. In contrast, a Brønsted acid donates them. Such acid-base reactions can sometimes be very vigorous but also

compound. To some extent, the properties of nitrogen may indeed be interesting for biochemistry. Therefore, biology ultimately uses nitrogen in the very important amino acids. These are the building blocks of proteins. However, the task of forming the backbone of biochemical molecules exceeds the capabilities of nitrogen. In amino acids or proteins built from them, nitrogen is indeed used. However, only as one of several elements.

Carbon, on the other hand, neither has the basic properties of nitrogen compounds nor are long chains of carbon atoms chemically unstable. Moreover, the tetravalency of carbon is a great advantage. Even if complex molecules could already be built in principle with trivalent molecules: With the ability to be bonded to up to four other atoms simultaneously, carbon offers many more possibilities for building complex molecules. That is why we say that terrestrial life is based on carbon. And for this reason, it is often speculated that life could also be based on silicon.

Silicon is located directly below carbon in the periodic table of elements. In the periodic table, the principle applies that elements that are below each other form a group. The elements of a group generally have similar chemical properties. Accordingly, carbon and silicon are similar in some essential points. This includes, among other things, the number of bonds they can form simultaneously. Silicon is also tetravalent and can form the so-called silanes analogously to hydrocarbons. Silanes are chains of silicon atoms with hydrogen atoms attached laterally. They thus correspond to alkanes as the simplest form of hydrocarbons. The carbon atoms are simply replaced by silicon atoms. If one were to replace one of these hydrogen atoms with a silicon atom, the chain could branch. If one were to replace a hydrogen atom with a completely different atom (or a group of atoms), the molecule could be functionalized. This means that one could give it various chemical properties, which is an essential prerequisite for a functioning biochemistry. This is the reason why speculation about silicon-based life initially makes quite a bit of sense.

But what challenges would silicon-based life forms actually face? One of the reasons why carbon is so well-suited as the basic material of biochemistry is the high stability of the bond between two carbon atoms. Carbon compounds simply do not break down easily. The bond between two silicon atoms, on the other hand, is significantly weaker. Accordingly, it would be even more important for a silicon-based life form than for us carbon-based life forms to have conditions that do not destroy their biochemistry. An important point in this context is temperature. Simply put: the higher the temperature, the more unstable silanes and their derived compounds are. Therefore, silicon-based life forms would likely develop primarily

---

play a significant role in biochemistry. It would be of little help if the backbone of all molecules were basic. This would cause the pH value inside living beings to rise enormously (i.e., the water would also become basic) and all acids would be neutralized. As a result, all acid-base reactions would be practically unusable for biochemistry because all Brønsted acids would have already given up their protons.

in a cold environment. Horta or the somewhat later discovered Excalbians, with whom we will deal in the next chapter and who are extremely hot, would therefore have enormous problems to contend with. Silicon-based life does not appreciate high temperatures.

However, low temperatures have certain disadvantages for life. A chemical problem in a cold environment is the so-called reaction kinetics. This describes how quickly a chemical reaction proceeds. The higher the temperature, the faster reactions occur. To reach (and maintain) a certain temperature level, mammals constantly burn fats and carbohydrates, their energy carriers, even if they do not actually need the corresponding energy at that moment. This is an enormous waste. But it still makes sense because it allows them to keep their body temperature constantly at a high level. Because chemical reactions occur faster at high temperatures, warm-blooded animals can carry out chemical reactions very quickly when needed. Why this is important becomes clear when you consider that the provision of energy for hunting, fleeing, or other physical exertions is provided by chemical reactions. If the temperature is low, the organism can only perform at a low level. Therefore, animals generally try to keep their body temperature high. Just not so high that the molecules of their biochemistry start to decompose. That is why the human body temperature is 37 °C. This is the highest possible temperature at which no damage occurs due to high temperature. At slightly higher temperatures, most organic molecules do not yet decompose. Nevertheless, individual (biochemically important) classes of substances, such as enzymes, already change and begin to lose their function. For silicon-based life forms, the maximum body temperature would be significantly lower. Accordingly, their biochemistry could still function. However, due to low temperatures, it would only function very slowly. Large physical performances, like the speed with which Horta moves, would be considerably more difficult for them than for us. Therefore, silicon-based life forms would probably hardly develop beyond the state of single-celled organisms.

The chemical stability of silicon-based biomolecules would not only be a problem with regard to temperature. Although silanes, the silicon equivalents to hydrocarbons mentioned above, generally withstand contact with water and mild acids without problems, as soon as the water becomes basic, they react violently and transform into solid silicates and hydrogen. Moreover, silicon-based life forms could not simply coexist with us in the same atmosphere. Silanes have another property related to their stability: they are pyrophoric. This means that they react violently with the oxygen in the air at room temperature. Pure silanes would actually start to burn. As we have seen above, all living beings are structured in such a way that they consist of small pouches (the cell membranes) containing a liquid (in our case, water). This liquid would likely prevent silicon-based life forms from immediately bursting into flames upon contact with air. Nevertheless, extraterrestrial life would face massive disadvantages if it replaced carbon with silicon.

What if extraterrestrial life forms were not only based on silicon but also had a completely different biochemistry? This is where it gets very speculative. The chemistry we will discuss next differs even more from any biochemistry in known living beings. But let's just speculate.

The weak point of silanes and the molecules derived from them was that the bond between two silicon atoms is too weak. Therefore, the molecules tend to be unstable. At this point, a widespread translation error could help. When English-language works refer to life forms based on silicon, German translations often speak of "Silikon". The German term Silikon does not refer to the chemical element silicon but to a group of plastics that include silicon in their chemical structure. The translation error becomes understandable when you know that silicon in English is called *"silicon"*. The plastic "Silikon" is called *"silicone"* in English. A small but not irrelevant difference.[3] Unintentionally, this translation error could point to a possible solution to the stability problem.

Let's take a closer look at silicones. Their basic structure is based on the class of substances known as siloxanes. These do not simply consist, as with silanes, of a series of interconnected silicon atoms, but always alternate between an oxygen atom and a silicon atom. Since the bond between oxygen and silicon is much more stable than between silicon and silicon, siloxanes are significantly more stable than silanes. Silicones, in turn, consist of siloxanes with alkyl groups attached to the sides. Alkyl groups are nothing more than parts of hydrocarbon molecules. In this way, you get a stable molecule that can branch out and to which all kinds of functional groups can be attached. In other words: Based on a silicone structure, you can essentially build everything that biochemistry needs. However, strictly speaking, life forms based on silicone would not be purely silicon-based life forms, but a hybrid form of silicon- and carbon-based.

There is one last major problem for silicon-based life forms. This concerns respiration. Respiration is important for living beings to provide energy. Although there are other chemical ways, such as fermentation, for living beings to gain energy, respiration is the only truly effective method. In respiration, carbon-based life forms convert organic compounds with atmospheric oxygen into water and carbon dioxide. Carbon dioxide, along with water, is a second substance whose special properties are essential for life. If the natural constants of the universe were just a little different, there would be considerable problems for life. Here are just two of them: First, there is the unusually high solubility of carbon dioxide in water. This allows carbon dioxide to be transported out of the body without gas bubbles forming in the cells, which would cause considerable damage. Secondly, although it is highly soluble in water, it is ultimately still a gas. This makes it very easy for an organism to release it into the environment.

This latter property, in particular, is a significant difference from what would happen in the respiration of silicon-based life forms. Analogous to carbon biochemistry, silanes or siloxanes would react with oxygen to form water and silicon dioxide. Carbon dioxide and silicon dioxide may sound very similar at first.

---

[3] This translation error also occasionally occurs in the German synchronization of Star Trek. In the 18th episode of the 1st TNG season, *"Home Soil,"* the German synchronization also mentions that the life forms on the planet Velara III are based on silicone.

However, there is a huge difference: Carbon dioxide is a gas. Silicon dioxide is, in the truest sense of the word, hard like a rock. Geologists usually call silicon dioxide quartz. The beautiful, glass-like crystals that you sometimes see are quartz: nothing more than silicon dioxide. A silicon-based life form should therefore avoid respiration. Otherwise, it must expect to turn itself into stone in no time.

One solution could be silicones again. As discussed, they consist of a siloxane chain with hydrocarbon residues attached. If silicon organisms could manage to specifically convert only these hydrocarbons to carbon dioxide during respiration and leave the silicon-containing backbone untouched, then respiration would be possible without turning themselves to stone.

In addition to true silicon-based life, which refers to life forms whose entire biochemistry is based on silicon, one could also imagine life forms that are carbon-based but also use silicon. If we think of our own bodies, they are, as an inorganic, silicon-based life form from *Star Trek: The Next Generation* puts it, essentially "ugly big bags, mostly filled with water" (the term "ugly" is in the eye of the beholder; according to the beauty ideals of the crystalline inhabitants of Velara III, this may well be true for humans). If this description were complete, humans would not be able to stand at all. To do so, the human (or other humanoid) readership of this book has bones. The bones also have a significant organic, i.e., carbon-based, component. However, they largely consist of calcium phosphate. This inorganic component makes the bones really hard. But there is no necessity for living beings to use calcium phosphate for this purpose. In the vast expanses of space, higher life forms could have developed whose bones consist of silicates. In principle, this even exists on Earth, albeit only in much simpler life forms. The so-called *diatoms* are tiny single-celled organisms that have a shell made of silicon dioxide. Although their actual biochemistry is (like that of humans) based on carbon, they at least use silicon for part of their biological functions. It therefore does not seem entirely far-fetched that silicon plays a role in biochemistry somewhere in the universe. There might be something to Horta after all.

> **A few more details**
>
> We are currently speculating about whether the biochemistry of extraterrestrial life forms could be based on elements other than carbon. Then we should at least be thorough enough to consider whether there might be other candidates besides the often-discussed silicon. Earlier in the text, I pointed out that elements that are in the same group in the periodic table are generally very similar chemically. This is the main reason why silicon is repeatedly brought up as a candidate. However, the periodic table continues below silicon. At least theoretically, one could imagine something like a biochemistry with the next element. Directly below silicon is the element germanium.
>
> The existence of this element was predicted by the Russian chemist Dmitri Mendeleev as early as 1871 from the structure of the periodic table, without anyone having isolated it at that time. In 1885, the German chemist Clemens

Winkler actually found it in a newly discovered mineral. It was eventually named after Winkler's homeland: Germany.

This germanium forms a group of substances called germanes, analogous to hydrocarbons or silanes. A germane molecule is a chain of germanium atoms, each with two hydrogen atoms attached laterally (or three hydrogen atoms on the terminal germanium atoms). In principle, one could again imagine a biochemistry based on this. However, the same difficulties as with silicon-based biochemistry arise, only much more pronounced. Germanes are chronically unstable, and in contact with air, solids are formed that do not melt even at 1000 °C. Therefore, germanium is probably not a serious candidate for non-carbon-based life. ◄

1.2 Very Hot Extraterrestrials

The Excalbians are a species remarkable in various ways. It starts on a cultural level. The concept of good and evil is completely unknown to them. They are apparently very fascinated by it during their first contact with the Federation. Therefore, they conduct an experiment that we can observe in the 22nd episode of the 3rd TOS season, *"The Savage Curtain"*. Their approach is to set up two teams. A total of eight people, consisting of members of the Enterprise crew and representations of historical figures, compete against each other in battle. One of the teams represents good. This team includes Captain Kirk, Mr. Spock, Abraham Lincoln, and the Vulcan philosopher Surak. The second team represents evil. It includes four correspondingly shady characters from Earth and extraterrestrial history. This team includes mass murderers like Colonel Green from Earth's history of the (from the perspective of the scriptwriter at the time, still far in the future) 21st century. But also personalities like Genghis Khan, who, at least in Mongolia, is considered more of a hero than an embodiment of evil, or Kahless, who (as one can learn in later Star Trek series) is practically the epitome of good for the Klingons. One begins to sense how much the assessment of good and evil lies in the eye of the beholder. Anyway, with the help of these two teams, the Excalbians want to determine what is actually better: good or evil.

The Excalbians do not really become wise from their experiment. This is most likely because their experiment has fundamental scientific weaknesses. On the one hand, the measurement methodology is probably only limitedly suitable for answering the actual question. Because what does the outcome of a fight to the death say? With high probability, the side to which the stronger and smarter fighters are assigned simply wins. The superiority of good or evil has at most a secondary influence on the outcome of the fight. On the other hand, the Excalbians are dissatisfied with the result for another reason. The good use similar methods in the fight as the evil. They cannot really see the difference and eventually have to release the Enterprise back into freedom without having really progressed with their question.

1.2 Very Hot Extraterrestrials

Besides their unusual misunderstanding of the concept of good and evil, the Excalbians are also extremely remarkable from a chemical perspective. On the one hand, they are chemically based on silicon. Therefore, when Scotty beams up the image of Abraham Lincoln created by them, he feels like he is beaming up a rock. We have already discussed the difficulties of silicon as a basis for life and the necessity of low temperatures for silicon-based life. This should not be the topic again here. On the other hand, the Excalbians are very hot. When trying to touch them, Captain Kirk first burns his hands properly.

As we have already established, silanes can obviously not form the basis of their biochemistry. At several hundred degrees Celsius, they would decompose in no time. In an oxygen-containing atmosphere, silicon would inevitably form silicon dioxide at these temperatures. Since Scotty had the impression when beaming that he was disassembling and transporting a rock into its atoms, silicon dioxide seems to make up a considerable part of the Excalbian organism. However, silicon dioxide cannot be the main component. Mineralogists call it quartz. If you are essentially made of quartz, then you are quite solid and quite restricted in your movements. The Excalbians do not make an overly agile impression. That would somehow fit with silicon dioxide. Even if the Excalbians may be several hundred degrees hot, quartz only melts at 1713°C. Not even the aliens from the planet Excalbia are that hot. Since the silicon dioxide in their bodies is both solid and chemically quite inert, it is not really suitable as a carrier of their biochemistry. It may possibly form the basic material of their bones. If their bones consist exclusively of silicon dioxide, then that might explain their slow movements.

Bones made of quartz would indeed be very hard. At the same time, however, they would also be quite brittle. Silicon dioxide possesses virtually no elasticity. Therefore, objects made of silicon dioxide break very easily. Just think of glass. The main component of most glasses is silicon dioxide. If you drop a drinking glass on the floor, the likelihood is quite high that it will break. If you drop a plastic cup on the floor, it usually survives unscathed. So, hardness is not everything. A certain flexibility is often quite helpful. Inorganic silicon dioxide, whether in the form of glass, quartz, or another mineral, is very hard. However, it almost completely lacks flexibility. Organic plastics, on the other hand, are relatively soft. They can be bent somewhat without breaking immediately.

The malleability of plastic can be chemically explained by the fact that plastic consists of a large number of molecules. These are quite large. There is a strong chemical bond between the atoms in the molecule. However, there is only a relatively weak attraction between the molecules. This is referred to as van der Waals forces. If you shift two molecules of the plastic against each other, you only have to overcome these relatively weak bonds. Therefore, plastic is quite soft. Subsequently, the same van der Waals forces exist between the molecules in the new position. In silicon dioxide, the entire solid (greatly simplified) consists of a single molecule. You cannot deform it without breaking relatively strong chemical bonds. That is why quartz is so hard. However, once you have shifted the atoms against each other, the chemical bonds are broken, and new chemical bonds do not form so easily. Therefore, when the chemical bonds break, the entire solid breaks.[4] Silicon dioxide is therefore quite brittle. If the bones of the Excalbians actually consist of silicon dioxide, then they should move very carefully because their bones would be very hard but also prone to brittle fracture. Additionally, they probably do not heal particularly well if they break.

For this reason, our bones are not simply made of lime. Inorganic calcium salts give the bone its hardness, but there is also a lot of organic tissue in the bones. This gives the bone a certain flexibility. Therefore, it is not as prone to fractures. And if a fracture does occur, the blood-supplied organic tissue can contribute to the healing of the broken bone. For example, calcium phosphate is transported into the bone through the blood. This not only hardens them during childhood but also allows the healing process in the event of a bone fracture. Silicon dioxide, on the other hand, is almost insoluble in water (and thus in blood). Excalbians would therefore have a problem if they wanted to heal a bone fracture. It is therefore very sensible of them to move very carefully.

However, the transport of silicon dioxide through the blood of the Excalbians is probably only one of their many problems. If their bodies are actually several hundred degrees hot, then they have to deal with entirely different challenges. The difficulties start much more fundamentally. Namely, with the state of aggregation of the blood. The blood (or another body fluid that takes over its function) must be liquid. A solid cannot naturally fulfill the task. But what about a gas?

A gas could theoretically flow through the veins by an extraterrestrial heart. If the blood of the Excalbians is based on water, this would become necessary because water evaporates (at a pressure of 1 bar) at 100 °C. While vapors might be conducted through Excalbian veins, it cannot really take over the function of blood. Because blood is primarily supposed to transport chemical substances through the body. In the case of oxygen, which is supposed to reach the cells from the lungs, this would still be quite simple. In the case of carbon dioxide, which is transported by the blood from the cells to the lungs, this is also not a problem. However, it becomes difficult with almost everything else. Energy carriers, vitamins, proteins (or whatever takes over these functions in silicon-based Excalbians) dissolve more or less well in water. At least as long as it is liquid. When it is evaporated, it becomes more difficult. How the Excalbian version of proteins is supposed to be transported by gaseous blood is completely unclear. It would be similarly difficult with energy carriers like sugar. Sugars are well soluble in water due to the numerous hydroxyl groups in the molecules. However, it is hardly possible to evaporate them without decomposing them. If you simply leave out the hydroxyl groups in the sugar molecules, evaporation works much better. Sugar

---

[4] The molecules of organic plastics can also be linked by chemical bonds, significantly increasing their hardness. These are then referred to as thermosets instead of thermoplastics. Thermosets are considerably harder and do not melt (eventually, they decompose chemically if the temperature becomes too high). However, the higher hardness is not only bought at the cost of poorer processability but also with a certain tendency to brittle fracture.

1.2 Very Hot Extraterrestrials

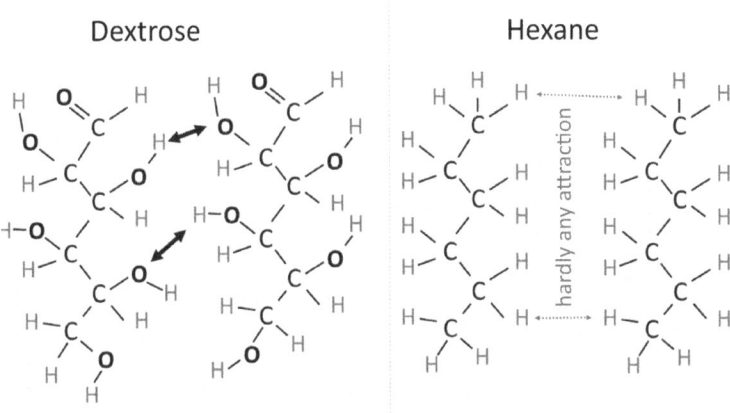

Fig. 1.1 Two molecules each of glucose (dextrose) and hexane and visualization of the attractive forces between the molecules; strong hydrogen bonds occur between the hydroxyl groups in glucose, while there are only comparatively weak van der Waals bonds between the hexane molecules (In reality, the carbon chain of glucose is usually closed into a ring; for simplicity, only the open-chain form is considered here)

molecules without hydroxyl groups would simply be hydrocarbon chains. Such substances are called alkanes. These alkanes are hardly soluble in water. For this reason, our organism can do very little with alkanes. However, if you do not have liquid blood, this does not really matter. Alkane vapors could then flow through the veins of the Excalbians.

Why is it so difficult to vaporize sugar? And why does the corresponding alkane (i.e., the "sugar" molecule without hydroxyl groups) vaporize so much more easily? When we compare glucose with hexane, we find that, in principle, it is essentially the same compound. In both cases, it involves six carbon atoms in a row (Fig. 1.1).[5] The difference is that in hexane (as in all alkanes), two hydrogen atoms are attached to each carbon atom (or three to the terminal carbons in the chain). In glucose, however, some of the hydrogen atoms are replaced by hydroxyl groups. A hydroxyl group consists of one oxygen and one hydrogen atom. Since the oxygen atom attracts electrons much more strongly than the hydrogen atom, a slightly negative charge results at the oxygen atom. Because the electron concentration at the hydrogen atom is somewhat reduced, a positive charge results there. The overall molecule is electrically neutral. However, the charge is unevenly distributed within the molecule. Opposite charges are known to attract each other.

---

[5] In glucose, the carbon atoms are usually not arranged in a long row but form a ring. Cyclohexane or methylcyclopentane would therefore be somewhat more accurate analogs to glucose. For simplicity's sake, we will assume here that it is just a long, straight chain. For understanding the effect we are discussing, it does not matter whether the carbon atoms form a ring or a linear chain.

Therefore, the negative oxygen atoms attract the positive hydrogen atoms in the hydroxyl groups of other sugar molecules. This effect is called hydrogen bonding. This effect does not occur in hexane. Therefore, the attractive forces between hexane molecules are significantly weaker. When the molecules attract each other less strongly, they are easier to separate. This is exactly what happens during vaporization. Therefore, an alkane like hexane vaporizes much more easily than sugar with the same number of carbon atoms in the molecule.[6]

Why have we examined this aspect so closely here? After all, this chapter is supposed to be about living beings like the Excalbians, who have a very high body temperature. The hydroxyl groups in sugars are precisely the crucial point. Therefore, the Excalbians cannot use sugars (or other carbohydrates) as energy carriers. The blood in their veins would be gaseous. Alkanes like hexane, on the other hand, would be quite suitable. They can be vaporized without decomposition and thus transported in the Excalbian body. Alkanes are already very well known as chemical energy carriers today. However, not from biology, but from combustion technology. Gasoline is ultimately nothing more than a mixture of alkanes. The energy content per kilogram of alkane is even significantly higher than that per kilogram of sugar.[7] As we will see a little later, it is very important for the Excalbians to have an energy-rich fuel as an energy source.

First, however, we want to stay with the aggregate state of Excalbian blood. Wouldn't there be a way to have liquid blood at several hundred degrees Celsius? After all, blood must not only be an energy carrier but also transport all sorts of things through the body. For this, a liquid is simply better suited than a gas. To keep a liquid liquid at higher temperatures, there are even three options (which can also be combined):

1. You can increase the pressure.
2. You can salt the water.
3. You can replace water with something else.

Option one actually occurs in nature. On Earth, there are organisms that live at very high temperatures. These are called thermophilic organisms. Many of these microorganisms live at temperatures above 40 °C. Some live at more than 70 °C. And some even at over 100 °C. The latter, however, only exist in the deep sea. Because only there is the pressure high enough to prevent the water from

---

[6] Hydrogen bonds can be formed not only with hydroxyl groups in other sugar molecules. They are also formed with (chemically similar) water molecules. This is why water is practically attracted to sugar, which is why its solubility in water is very good. Hexane, on the other hand, does not form hydrogen bonds with water and therefore dissolves poorly in it.

[7] A problem with regard to energy would be the volume. One liter of liquid gasoline contains a lot of energy. However, when vaporized, it expands significantly. The calorific value of the gasoline does not change. However, the space requirement increases significantly. Since hexane already boils at 69 °C, the Excalbians would have to constantly take in new alkanes, as they can hardly store reserves in their bodies due to space constraints.

evaporating. At normal atmospheric pressure, the cells would simply be torn apart by evaporation. Near deep-sea volcanoes, temperatures are sometimes above 100 °C, and the water is still liquid. Hyperthermophilic organisms live here. So it is possible to keep water liquid in living beings even at high temperatures. How high does the pressure actually have to be for this?

The higher the temperature, the higher the pressure must be. With rising temperature, the required pressure increases approximately exponentially. At 150 °C, a pressure of about 5 bar would be necessary. At a depth of 40 meters underwater, such a pressure is reached on Earth. At 250 °C, at least 38 bar is already needed. This corresponds to a diving depth of about 370 meters. Since Captain Kirk and Mr. Spock were not crushed on Excalbia, the pressure there does not seem to be quite as high. On the other hand, the internal pressure of the Excalbians' bodies cannot be much higher than the ambient pressure. Otherwise, they would simply explode. Therefore, increased pressure is probably not the explanation for how the Excalbians can be so hot.

Moreover, there is an upper limit for keeping liquids under increased pressure. Every chemical substance has a so-called critical point. Above this point, no condensation can be achieved by increasing pressure. In the case of water, this critical point is at 374 °C and 221 bar. At the critical point, the steam is compressed so much that it is no longer distinguishable from the liquid. Condensation, as we know it, is therefore no longer possible. Beyond 374 °C, liquid water is definitely out of the question, no matter how high the pressure is.

The second option would be to add salt to the water. Our blood contains a certain amount of salt. But that does not mean that the blood of the Excalbians could not have a much higher salt concentration. Adding salt to water raises the boiling point. Chemically, this can be explained by two effects. On the one hand, the ions of the salt attract the molecules of the water. This hinders their transition to the vapor phase. On the other hand, the salt lowers the concentration of the water. Simply put, there are fewer water molecules on the surface, and thus fewer water molecules can transition to the vapor phase. The apparent boiling point therefore rises, and the blood would remain liquid at slightly higher temperatures.

The third option would be to use blood that is not based on water at all. Water is not only suitable as the main component of blood and all cells because it is liquid under Earth's conditions. Water also has various properties that are very important for biochemistry. But who knows exactly how Excalbian biochemistry works. Therefore, we can at least theoretically consider many substances that could take on the role of water. Sulfur, for example, remains liquid up to well over 400 °C. Perhaps liquid sulfur transports nutrients through the veins of the Excalbians? Alternatively, chemistry offers a multitude of other substances that would be liquid at the corresponding temperature. However, neither sulfur nor any of these other substances offer the chemical properties that make water so excellent for sustaining life.

Keeping the blood liquid is not the only challenge that high temperatures pose for the Excalbians. Another problem is the temperature difference to the environment. Unless the environment is also very hot, they give off a lot of heat to the surroundings. We do not know exactly how the Excalbians normally live. When they meet Captain Kirk and Mr. Spock, however, the ambient temperature does not seem to be much above 20 °C. To avoid cooling down very quickly, they must convert enormous amounts of energy carriers. What sounds like an excellent diet program would, in practice, lead to starvation very quickly. This could be one of the reasons why the Excalbians have rather shapeless, round bodies. The sphere is the shape that has the least surface area for a given volume. Therefore, a spherical organism loses less heat to the environment than an organism of the same size with a different shape. The closer the body approximates the spherical shape, the less heat it gives off to the environment. The body shape of the Excalbians seems to take this problem into account. Nevertheless, they would still have to chemically convert very large amounts of energy carriers to maintain their body temperature. They would therefore rely on a very energy-rich energy carrier like alkanes.

A final problem is the stability of biochemical molecules. We have already seen that silanes decompose at high temperatures. This is a problem for silicon-based life forms. Ultimately, carbon-based life forms like us would have the same problem. Above 40 °C, our proteins begin to denature. This is known from boiling eggs. In this process, proteins coagulate, which corresponds to said denaturation. They can no longer fulfill their biological function. The enzymes stop working and eventually become irreparably damaged. As catalysts of biology, enzymes are supposed to make desired reactions possible and provide an advantage over undesired ones. Without functioning enzymes, many reactions may still occur in the organism. Unfortunately, they are not the ones the body actually needs. Therefore, the rule is: an increased body temperature has its advantages. But only up to a certain point. Then biochemistry begins to break down.

We have already mentioned thermophilic organisms above. There are apparently ways to stabilize proteins so that they are stable at higher temperatures. How exactly this works goes a bit too far here. If only because the biochemistry of silicon-based Excalbians is completely different anyway. In any case, they must biochemically exert considerable effort to prevent essential molecules in their bodies from decomposing.

However, high temperatures are not only disadvantageous. Extraterrestrial life forms that are over a hundred degrees hot would have a number of advantages.

One thing that came to my mind is disinfection. Bacteria should hardly be a problem for the Excalbians, as they would be roasted within seconds. However, after thinking about it for a while, I realized that this circumstance only prevents them from contracting an infectious disease when in contact with Captain Kirk. Pathogens that have developed in the same environment as the Excalbians are likely to have similar biochemistry and therefore cope quite well with the high temperature.

A high temperature, however, has other advantages. One advantage would be that diffusion is greatly accelerated. In organisms, many substances are simply transported by diffusing. This means that the molecules move through another substance without external influence. One can imagine it as all molecules moving. In solids, they only oscillate around a predetermined place. In gases and liquids, they move quite chaotically. The molecules of a substance dissolved in water are therefore constantly being rammed by water molecules. As a result, the dissolved molecules themselves start to move. This movement is initially chaotic and constantly changes direction (namely whenever the next collision occurs). On balance, however, it happens more often that a molecule moves from an area of high concentration to an area of low concentration. How could it ultimately be otherwise: In an area of low concentration, there are hardly any molecules that can leave it. Consequently, diffusion is driven by concentration differences. Substances thus diffuse from areas of high concentration to areas of low concentration.

The speed of diffusion depends on three factors: the distance, the concentration difference, and the diffusion coefficient.

1. The shorter the distance to be covered by diffusion, the faster it goes.
2. The higher the difference in concentration, the more effectively the molecules are transported.
3. The higher the diffusion coefficient, the better diffusion works.

And the higher the temperature, the higher the diffusion coefficient. Simplified, one can imagine that as the temperature rises, the molecules move faster. The water molecules thus collide more often and more violently with the dissolved molecules. Accordingly, other molecules are transported faster through the substance by diffusion. A fast transport of needed substances in the body is very helpful for many reasons. We will examine this in more detail later.

The most important advantage, however, is the reaction speed. Chemical reactions proceed faster the higher the temperature. That is on of the reasons why mammals maintain their body temperature at about 37 °C. This is the highest possible temperature at which there are no problems due to the decomposition of enzymes and other biomolecules. However, if an organism manages to prevent this decomposition, it can further increase its body temperature. As a result, the chemical reactions inside it proceed significantly faster. Even if one enzyme or another were to decompose, it might be bearable. After all, enzymes serve to accelerate chemical reactions in the organism. Possibly, the Excalbians do not need enzymes at all due to their high body temperature.[8]

---

[8] At least the importance of enzymes for the mere acceleration of reactions is less important at high body temperatures. However, by accelerating the desired reactions, enzymes also give them an advantage over the undesired ones. In this way, enzymes contribute to ensuring that the right reactions occur in the organism and indirectly suppress undesired reactions because they are too slow compared to the desired ones.

1.3 Life without a Body

Bodies can be a very annoying matter. You constantly have to be careful not to bump and injure yourself. Bodies cannot withstand particularly high temperatures. Otherwise, the proteins decompose. On the other hand, bodies do not like temperatures that are too low. Otherwise, too many nutrients have to be burned, and chemical reactions proceed more slowly. Bodies break down when they come into contact with the wrong chemicals. Bodies get sick when they are infected with viruses or bacteria. Bodies are subject to gravity. Bodies are prevented by solid matter from passing through walls. All of this is extremely annoying and severely limits the possibilities that one has as a physical being.

In the vast expanses of space, there are obviously—as we learn from Star Trek—some species that have evolved to the point where bodies no longer constrain them. Our heroes from Star Trek, for example, encounter a whole series of bodiless life forms. Just think of the nebula-like creature that the Enterprise under Captain Kirk encounters in the 18th episode of the 2nd TOS season, *"Obsession,"* or the Companion that the away team of the Enterprise meets in the 2nd episode of the 2nd TOS season, *"Metamorphosis,"* not to mention the bodiless creature that feeds on hate and therefore makes Kirk and his crew fight the Klingons in the 11th episode of the 3rd TOS season, *"Day of the Dove."* Later Starfleet crews also encounter bodiless life forms. Think of Nagilum, with whom Captain Picard and the crew of the Enterprise-D have to deal in the 2nd episode of the 2nd TNG season, *"Where Silence Has Lease,"* or the bodiless being from the 21st episode of the 4th DS9 season, *"The Muse,"* which inspires Jake Sisko to achieve his highest literary accomplishments. Not to mention the photonic life forms that cause problems for the crew of the Voyager in the 12th episode of the 5th VOY season, *"Bride of Chaotica,"* or the Organians, who appear not only on the Enterprise under Captain Archer in the 11th episode of the 4th ENT season, *"Observer Effect."*

All that is just a small selection. One could continue the list further. Bodiless life forms apparently fascinate people even more than life forms based on silicon. Reason enough to take a look at bodiless organisms. We won't even start with life forms that consist only of pure energy. That would be so far beyond anything we know that we couldn't conceive of it anyway. Therefore, we will briefly look at the simplest basic form of bodiless organisms: Gaseous life forms.

What challenges would such a mist being face? What could its biochemistry look like? In the 6th episode of the 1st VOY season, *"The Cloud,"* we learn that the cosmic cloud, which turns out to be a life form, consists of hydrogen, helium, and hydroxyl radicals. Additionally, there seem to be Omicron particles inside it.

What Omicron particles are cannot be answered according to the current state of science. There are elementary particles designated with the Greek letter Omega. However, an Omicron particle is not yet known. It can be assumed that these Omicron particles play a significant role in actually making the cloud a life form. Because the other substances are not really suitable for maintaining biochemistry. Hydrogen and helium are indeed common in the universe. However, nothing can be done chemically with helium. If you cannot form a bond with any other atom,

1.3 Life without a Body

you cannot contribute to biochemistry. This is the case with helium, and therefore this element is completely useless for any biochemistry. Not just for the biochemistry we know on Earth. Hydrogen can indeed form chemical bonds. On the other hand, it can only participate in them. It cannot function as a basic substance. As we have already seen, hydrogen is monovalent. If you can only form a bond with a single other atom, you can only form diatomic molecules (without the help of another element like carbon). You still can't really do anything with that.

A living being might be able to do a bit more with the hydroxyl radicals. The term hydroxyl indicates that it is a compound of hydrogen and oxygen. Oxygen is at least divalent. Apart from the fact that long chains of oxygen atoms are unstable, no complex molecules can be built from them. The possibility of branching the chain is missing. The fact that the hydroxyls are radicals also indicates that we are dealing with "incomplete" molecules of oxygen and hydrogen. These are very reactive. Therefore, they are called radicals. In simple terms, a bond in radicals leads to nowhere. At this point, the molecule is very eager to form a bond because it only has a half bond in a certain way. This is where the high reactivity comes from. Normally, you want to keep radicals away from the body. Due to their high reactivity, they easily "destroy" other molecules. For the extraterrestrial gas organism, reactions that these radicals carry out could theoretically provide energy. However, a proper biochemistry still does not work with that. Quite apart from the fact that the scanners of the Voyager obviously do not find significant amounts of the corresponding reaction partners or products. If the conditions are such that the hydroxyl radicals can react properly, then substances like water or hydrogen peroxide should be found in the cloud.[9]

It looks similar with the aforementioned Companion, a bodiless life form that Captain Kirk encounters in the 2nd episode of the 2nd TOS season, *"Metamorphosis,"* which consists only of energy and ionized hydrogen. The high amount of energy might explain why the hydrogen is ionized. Nevertheless, the Companion would still only be a cloud of hydrogen plasma[10]. It would thus be

---

[9] Hydroxyl radicals each have a single electron on the oxygen atom. That is the bond "to nowhere." When two radicals meet, they link up so that the two "half bonds" unite into a "whole bond." In the case of two hydroxyl radicals, a molecule would form in which two oxygen atoms are linked (this type of molecule is called peroxide). Since each oxygen atom still has a hydrogen atom attached, the resulting compound would be hydrogen peroxide. Like all peroxides, however, this is rather unstable, so its decomposition would not be surprising. Among other things, water would be produced during the decomposition.

[10] The term plasma refers to a gas in which the atoms have given up at least some of their electrons. In a certain way, the gas then consists of two types of particles: anions and electrons (plus often a certain part of non-ionized atoms or molecules). The anions are the atomic nuclei that remain when some of the electrons have been given up. In the extreme case, all electrons have been given up. Then the atomic nucleus is only the atomic core. Since a hydrogen atom has only one electron, hydrogen plasma necessarily consists of atomic nuclei. A hydrogen atom cannot be partially ionized. A plasma can be created, for example, by high temperatures, but also by radiation.

quite electrically conductive. Nevertheless: That still does not result in biochemistry. To live, an organism is needed. It must be able to actively do something. Otherwise, it is not a living organism. It needs molecules that perform functions in the organism. For example, it needs proteins or similar substances to build structures. Without any structure, no process can really take place. Even if these structures are not proper organs, there should at least be formations like cells. Without proteins (or something similar), such structures simply cannot be built. Additionally, something like nucleic acids (i.e., a kind of DNA) is needed to store genetic information. Otherwise, the organism could not reproduce and realistically could not synthesize biomolecules either. All these are complex molecules. With only hydrogen and maybe some oxygen, you can only be a cloud. Life does not work that way.

A serious chemical challenge is also likely to be the reaction rate. For a chemical reaction to occur, the corresponding molecules must collide. This happens more frequently the smaller the distances between the molecules are. Terrestrial biochemistry takes place almost entirely in aqueous solution. In other words: in a liquid. In liquids, the molecules are packed almost as closely as in a solid. The arrangement is just a bit more chaotic. Due to the small distance, collisions occur very frequently. Frequent collisions can often lead to reactions between the molecules. The reaction rate is correspondingly high. In a gas, the average distance between the molecules is much larger. Therefore, collisions occur less frequently. If the gas is also in space, then the pressure is very low. Accordingly, the average distances between the molecules are very large (or better said from a molecular perspective: gigantic). This, in turn, affects the frequency of collisions. Chemical reactions therefore proceed slowly. A gaseous life form would therefore hardly be able to quickly provide a lot of energy, for example. Thus, its performance is very low.

Another problem for bodiless life forms would be the cohesion of the molecules. What prevents the gas molecules from dispersing into space? Or, if the gas life form is on a planet, what prevents it from mixing with the planet's atmosphere? To keep their components together, cells of organisms on Earth are surrounded by membranes. However, membranes are solid structures. Although they are only micrometers thick and highly deformable, they may not seem particularly solid to us. Nonetheless, a membrane is a solid. This solid limits the body of a biological organism. Thanks to cell membranes, our cells are held together. As a result, the cell components that are supposed to stay together actually stay together. The fascination of bodiless life forms may lie in the fact that they do not have a limited body. But if they simply continue to disperse into the environment, they cannot survive for long. They simply dissolve—a notion that, on the other hand, is not particularly fascinating.

A way out of this dilemma could be gravity. At least a huge cosmic nebula like the one just discussed could simply be held together by its own gravity. If the nebula being is really several astronomical units large and its density is not too low, then it has a certain mass. Mass, in turn, leads to attraction in the form of gravity. This gravitational field could hold the nebula being together. Curiously, however, this nebula being is the only bodiless life form in Star Trek that seems to possess something like a cell membrane. The crew of the Voyager only notices this after they have damaged it by simply flying through it, leaving a large hole in it. Fortunately for the nebula being, this hole is caused by a Starfleet ship. The Voyager does not simply fly on. First, a way is found to "sew" the wound with a novel form of space surgery.

1.4 Crossing the Threshold

An episode that raises scientific questions like few others is the 15th episode of the 2nd VOY season. It bears the title: „*Threshold*". On its journey home from the Delta Quadrant, the Voyager discovered a completely new form of dilithium. What dilithium actually is would undoubtedly be an incredibly exciting question from a chemical perspective. After all, it is mentioned countless times in Star Trek and plays a crucial role in the warp drive. Obviously, it cannot simply be synthesized but must be mined on foreign planets. If such synthesis is still not possible in the 24th century, it suggests that dilithium is an (as yet unknown to us) chemical element. How to imagine a new, previously unknown form of a known chemical element is another exciting question. A new isotope seems to be the only explanation that makes sense according to current knowledge. Then the atomic nucleus of this new form of dilithium would simply contain a few more or less neutrons. Its chemical properties would remain unchanged, but some physical properties could differ significantly.[11]

This new form of dilithium apparently possesses quite remarkable physical properties, as it enables the construction of a transwarp drive. The warp drive allows Starfleet ships to travel faster than light. Current human research and development are still far from this. According to the current state of knowledge, the theory of relativity tells us that this is not so easily possible. In the future, however, this problem will somehow be solved, allowing travel to foreign planets (otherwise, we would be dealing with a serious error in Star Trek, and we really want to dismiss that thought). For the Voyager, however, warp speed is still too slow. It is at the other end of the galaxy and would take 70 years to return to Earth even at the highest speed its warp drive can provide. The discovery of the foundations for transwarp travel is therefore of particular interest to the Voyager crew. Transwarp means infinite speed. One could occupy any point in the universe simultaneously

---

[11] We will come back to the topic of isotopes. Theoretically, one could also imagine that the new form of dilithium involves isomers. Isomerism is actually a term that plays a major role in chemistry but has nothing to do with the properties of the atoms themselves, rather with their arrangement in the molecule. However, the term isomers is also used in nuclear physics and refers to atomic nuclei that do not differ in the number of protons or neutrons but have different energy levels. Energetically excited states of atomic nuclei, even if they are very long-lived, usually revert to the ground state within nanoseconds. Gamma radiation is released in the process. But perhaps there is a stable isomer of dilithium in the Delta Quadrant.

or be home in the blink of an eye. No wonder, then, that every effort is made to develop the corresponding technology, which is achieved in the remarkably short time of just one month.

The possibility of infinite speed is certainly very fascinating from a physical standpoint, but it will not concern us further here. We also do not want to further discuss the question of how the human sensory apparatus is supposed to take in impressions from the entire universe at once and how the brain is supposed to process them. What happens to the pilot of the first transwarp flight is much more exciting biochemically.

Tom Paris is allowed to pilot the shuttle Cochrane, contrary to the medical advice of the ship's doctor. Although he loses consciousness during the flight, he seems to have otherwise coped well with the journey through the entire universe. At least as far as the first impression goes. However, not long after his return, his body begins to change at a rapid pace. He develops an allergy to water and can no longer breathe normal air. To still enable Tom Paris to breathe, the doctor replaces the oxygen in a small part of the sickbay with acidichloride.

The question of what kind of substance this is certainly raises puzzles. The word component "dichloride" suggests that the corresponding molecule contains two chlorine atoms. These are located on a molecular framework that is further specified by the syllable "aci-". According to the IUPAC rules for the nomenclature of chemical compounds, this component of the molecule cannot be identified. Some may alternatively assign the letter "d" to the word component "aci". Then one can read the first part of the word as "acid". However, the question arises as to what the "i" between "acid" and "chloride" is about. If one were to insert a second "d" into the word, it would result in a molecule based on some acid and two chlorine atoms. There are indeed acids that contain two chlorine atoms in the molecule (for example: dichloroacetic acid, which is used as a solvent and starting material for the synthesis of some other substances). What exactly the doctor uses to enable the mutating Tom Paris to breathe remains unclear.

The rapid changes that Tom Paris's body undergoes are eventually explained by the doctor as a natural process. Apparently, the lieutenant is undergoing normal evolutionary advancement. This is quite remarkable if one understands what evolution actually is. The theory of evolution states that organisms adapt to changed living conditions over time. The genetic information that leads to physical traits best suited to the environment prevails. There are two aspects to consider here.

The first point is that it is about adaptation to changed living conditions. Evolution does not aim towards a "higher" life form. Greater body size or intelligence can certainly prevail. But they do not have to. It depends on the environment what is "higher". If this changes, new evolutionary processes set in that lead in a different direction. The development from tiny single-celled organisms to huge dinosaurs had definite advantages. That is why it happened in the past. However, if the environmental conditions change at some point, the seemingly "higher life form" with its large body size suddenly becomes disadvantaged and suffers the same fate as the dinosaurs once did, or later the mammoth and other prehistoric giants. Therefore, there is no "higher" development in evolution, only "better adapted".[12] Depending on the environment in which the organisms live, the best adaptation can be something entirely different. Accordingly, there is no predetermined direction in which evolutionary development would proceed. This is somewhat hinted at the end of the episode. After Tom Paris overpowers Captain Janeway and undertakes a second transwarp flight with her, the two eventually strand on an alien planet. There they find themselves in a swampy jungle, where they do not develop into mega-intelligent superhumans. Instead, they become large amphibians that give birth to their young in an earth hole. A swampy jungle provides entirely different conditions than a highly technological spaceship. Optimal adaptation leads to a completely different result there.

The second point is the speed of evolution. Tom Paris transforms into a life form that differs greatly from conventional humanoids within a few days. Although bacteria can make adaptations to changed environmental conditions through evolutionary processes within remarkably short periods, there is a significant difference. The generation time of bacteria can be significantly less than an hour under certain circumstances. A few days then already constitute several generations. In contrast, Tom Paris undergoes exactly zero generational changes in these few days. Therefore, no evolution can take place.

Evolution is based on those who reproduce more being better adapted to the given environmental conditions.[13] Those who do not (or less) reproduce because they, for example, do not survive, will have their genes less widely spread in future generations. Therefore, the gene pool of a population approaches the genome of those individuals who reproduce the most. The resulting changes in genetically determined traits are called evolution. Without reproduction, there is consequently no evolutionary development. An individual cannot therefore undergo evolution within its own lifespan.

So what happens to Tom Paris? It clearly cannot be a classic evolution. For one, it can only be influenced to a limited extent (for example, by radioactivity, because it causes more mutations and consequently greater changes per generation, or by drastically changed environmental conditions, because a larger part of the population dies and only those who can better cope with these conditions due to their genetic constitution are left to reproduce). On the other hand, it is ultimately only

---

[12] Darwin's famous formulation "survival of the fittest" accordingly does not mean the selection (i.e., the superiority) of the strongest, but of the best adapted. The English word "fit" at this point has nothing to do with fitness in the sense of physical strength, but comes from "to fit" (to suit). If size and strength are what is best under the given environmental conditions, then the largest and strongest will prevail. If the conditions are different, then perhaps the smallest will prevail. Because under changed conditions, e.g., with food scarcity, size may be a real disadvantage because it entails a large need for food.

[13] Here lies a second, widespread misunderstanding about evolution. It is not primarily about surviving oneself, but about reproducing. Survival is important insofar as it is, in a way, a prerequisite for successful reproduction. Someone who has ten offspring but ultimately dies quite early is still evolutionarily more successful than someone who lives to an old age but has only one or two (surviving) offspring.

about a single individual and not a sequence of generations. This question is quite interesting biochemically.

Somehow, the journey at infinite speed seems to have triggered a process that permanently rewrites Tom's genetic material. Several possible causes that induce mutations are already known today. Radioactive radiation, for example, can lead to mutations. Or a whole range of chemical substances. Such substances are called mutagens. Examples of such mutagens are phenol or benzene. Since not even the medicine of the 24th century really understood this consequence of the transwarp flight, it is naturally difficult for us to say what was actually going on. But we can at least try to understand the problem itself.

For this, we first need to realize how DNA molecules are actually distributed in the body. The genetic material of a human is not stored in a central organ. Therefore, you cannot simply change it in this place and thereby change the genome of the entire human. Instead, there are multiple copies of the genetic information. And we are not talking about one or two backup copies here. The DNA molecules with the genetic information of a human are present in approximately 100 trillion copies. This is how many cells the body of an adult human consists of (plus or minus a few tens of trillions). In each of these cells (more precisely: in the respective cell nucleus) there is a copy of the DNA molecule. So if the transwarp flight triggers a change in a DNA molecule in one of these cells, it has more or less no impact. A changed genome in a single cell is completely insignificant. Due to the changed DNA, this cell may produce slightly different proteins. Its biochemistry is therefore no longer the same. Measured against the total number of body cells, however, this is completely irrelevant. It only becomes relevant if the change leads to the cell multiplying very strongly. The cell then becomes the origin of a tumor. However, the mutation of a single cell cannot cause significant changes in the appearance or abilities of a human.

The mutation would therefore have to affect very many cells simultaneously. For a human to experience a real transformation, essentially every cell in the body would have to mutate. And here lies the real challenge: Why should they all mutate in the same way? Actually, each cell would mutate in a different way. Let's take a look at the biochemical basics of mutations.

Mutations usually occur when an error happens while copying DNA. DNA is constantly being copied. Whenever a cell divides, it has to double its DNA to pass the full genetic information to the daughter cell. For this, the genetic information stored in the DNA molecule is essentially "transcribed" once. DNA molecules are known to form the famous double helix. Two molecular strands are twisted around each other. Each of these strands consists—simply put—of phosphoric acid molecules and deoxyribose molecules, which are alternately linked together. This creates a long chain. Attached to each deoxyribose molecule is a nucleobase. There are four different ones: adenine, guanine, thymine, and cytosine.[14] The two

---

[14] At least this is the case with DNA. In RNA, which is used to convert genetic information into proteins, uracil is used instead of thymine. The two molecules are quite similar chemically. Thymine merely has an additional methyl group.

1.4 Crossing the Threshold

molecular strands of the double helix are not actually connected to form a single molecule. There is no covalent bond between them, but only very many hydrogen bonds. However, this does not need to concern us further.

The two strands of the double helix complement each other. This does not mean that they are identical. In a way, the counterpart is always found in the other strand. If there is an adenine molecule in one strand, there is a thymine molecule at the corresponding position in the other strand. If there is a guanine molecule in one strand, there is a cytosine molecule on the opposite side. There is always an adenine-thymine or guanine-cytosine pairing. When DNA is copied, the double helix is separated, and the appropriate nucleobases (along with the deoxyribose and phosphoric acid residue) attach to each strand: always adenine and thymine together and guanine and cytosine together. In this way, DNA molecules can double. The result is two identical molecules that exactly match the original molecule.

A mutation occurs when a "transcription error" happens. There can be many causes for this. Radioactive radiation, for example, can damage individual nucleobase molecules. As a result, they are overlooked during transcription. Mutagenic substances can alternatively insert themselves during transcription. This can also lead to a transcription error. Substances that are chemically very similar to nucleobases can be mistakenly read along. In all these cases, the copy is no longer identical to the template. Such a mutation usually has no significant effects. If the consequence is not uncontrolled proliferation (cancer), it is essentially irrelevant. In the worst case, the cell dies. With 100 trillion cells, this hardly matters to the body. Only if the mutation affects a germ cell (egg or sperm), does the mutation become relevant. Because then it can be passed on to the offspring. In the offspring, the mutation then appears in all cells of the body because they all originated from the parental germ cells. In this case, the mutation can influence the further development of evolution. However, most mutations should not be overestimated. Each person differs on average by about 50 mutations from their parents (in other words: we are all mutants!). Given the enormous extent of the human genome, this usually hardly matters. Only very few mutations actually cause a significant change in the characteristics of the offspring.

Now back to Tom Paris. His body is mutating, which may be caused by some kind of radiation during the transwarp flight. However, three things are noticeable in this process:

1. Why does the process only start after quite a while?
2. Why does the process proceed in an "evolutionarily directed" manner?
3. Why does it proceed the same way in all cells?

I have no answer to question 1. Radiation can hardly explain it. At most, a chemical that acts with a delay because, for example, it first has to penetrate the cell nuclei, which takes some time, might be a reason. But how could he have been administered this chemical during the transwarp flight? We have already discussed question 2 above. We understand the problem with question 3 when we recall what we have just learned about mutations. Mutations are random events. If two cells mutate, then two different daughter cells result. In the two cells, it is highly likely that completely different genes have mutated. And even if they were the same genes, it is still likely a different mutation of the same gene. And if 100 trillion cells mutate, then 100 trillion different daughter cells result. Therefore, the DNA of a human in all his cells cannot be rewritten by mutation. This is only possible during generational change. Children can have DNA that differs (slightly) from that of their parents as a result of mutations. If Tom Paris mutates as a whole after his transwarp flight, then we are dealing with a biochemical effect that is at least very unusual and cannot be explained with today's biochemical knowledge. Perhaps science in the 24th century will eventually find an explanation for it. Until then, we can continue to speculate about it.

Hydrogen and the Infinite Vastness

2.1 Breathing Hydrogen

In the 6th episode of the 2nd DS9 season, *"The 'Melora' Problem"*, a romance begins to develop. Dr. Julian Bashir falls in love (once again). A new crew member arrives on the space station: the Elaysian Ensign Melora Pazlar. The Elaysians come from a planet that seems to be much smaller than Earth, as the gravity is significantly lower than on other Class-M planets.[1]

If we recall the film footage from the late 1960s and early 1970s, the first Earth astronauts hopped around quite cheerfully on the Moon. Despite bulky spacesuits, they made impressive jumps on the lunar surface. This is because they had the muscles and skeleton of a species that evolved on Earth. And on Earth's surface, there is six times the gravity compared to the Moon. The Moon's gravity can hardly hold a human to the ground. The human body is almost overpowered for such an environment. Conversely, it is different for Elaysians. The Elaysians evolved on a planet with very low gravity. They are much better adapted to such an environment and can move more agilely there than most humanoids. However, on a Bajoran space station like Deep Space Nine, the artificial gravity is set to be similar to that on most Class-M planets. Its value is very similar to Earth's. As a result, the Elaysian Melora is confined to a wheelchair when she wants to move in the world of other humanoids.

---

[1] Alternatively, it is also conceivable that the density of the planet is significantly lower than that of Earth. The average density of Earth is about 5.5 t per cubic meter. A planet that is not primarily composed of iron would likely have a significant radiation problem, as it would lack the magnetic field to protect it from cosmic radiation. On the other hand, its density could be significantly lower. The density of quartz, which makes up a large part of Earth's outer crust, is only about 2.7 t per cubic meter. A quartz planet could theoretically be the same size as Earth but still have lower gravity.

© The Author(s), under exclusive license to Springer-Verlag GmbH, DE, part of Springer Nature 2025
K. Müller, *Chemistry and Science Fiction*,
https://doi.org/10.1007/978-3-662-70379-3_2

Dr. Bashir naturally tries to find a medical solution to this problem and develops a therapy that would allow Melora to adapt to our gravity. The catch: for medical reasons, she would then no longer be able to return to her world. The romance suffers from a classic long-distance relationship problem despite the actual physical proximity: the two lovers live in two different worlds, quite literally. Neither can really live reasonably in the other's world permanently. This naturally strains a relationship and becomes one of the main themes of the episode.

When Melora is on an exploratory flight with Jadzia Dax aboard the shuttle USS Orinoco in the Gamma Quadrant, the two women discuss exactly this problem of the relationship between Melora and Julian. To show that a relationship can work even across a difficult interspecies boundary, Dax tells her about a couple she once knew: one of the partners was a normal, oxygen-breathing humanoid. The second partner was a hydrogen-breathing Lothra. So, the two were a couple that couldn't even stay in the same room for long periods.[2]

The question of how to imagine hydrogen-breathing organisms becomes really exciting from a chemical perspective. Let's start by looking at the oxygen breathers we know, including us humans. Why do we actually breathe? It is primarily a question of efficiency. All animal life forms live by chemically converting energy-rich nutrients like sugar into lower-energy waste products. To do this, one does not necessarily have to breathe. Many microorganisms are capable of anaerobic living (the human body is also limitedly capable of this in situations of oxygen deficiency during intense physical exertion). One waste product that results from such anaerobic conversion is ethanol (also known as ethyl alcohol). This alcohol is not only very popular as a component of countless beverages. It is also used as a fuel.[3] If something can be used as a fuel, it must contain a lot of energy. In other words: if an organism converts sugar anaerobically into alcohol, much of the original energy is still in the waste product alcohol (which is biochemically a waste product). This portion of the energy is then not available for the organism's energy needs. This is where breathing comes into play.

A breathing organism does not convert sugars, fats, and the like into alcohol, but into water and carbon dioxide. This reaction is called oxidation. The term oxidation originates from the French chemist Antoine Laurent de Lavoisier. He

---

[2] Jadzia reports that the two could spend up to forty minutes in the same room thanks to intensive training. For a species other than humans, this might be realistic. However, an issue could be explosion protection. Even if neither of them breathes in and out in the other's atmosphere—to "breathe" there is a risk of explosion. If they want to talk, they inevitably have to exhale. This creates a mixture of air and hydrogen, also known as oxyhydrogen gas. Not to mention that the atmospheres would mix when entering the room. The explosion limits of hydrogen are much broader than those of other flammable gases. This means that even relatively low hydrogen concentrations in the air (or oxygen concentrations in hydrogen) are explosive. So, one can only hope that the two are non-smokers (although this vice does not seem to be very widespread in the future of Star Trek anyway).

[3] Often, not pure ethanol is used, but gasoline is mixed with a certain percentage of biologically derived ethanol. E10, for example, means that it is a gasoline with 10% ethanol content.

originally used it to refer only to the reaction of any substance with oxygen.[4] Over time, chemists have found that there are many reactions very similar to reactions with oxygen. These reactions were therefore included in the term oxidation. As a modern definition of the term oxidation, it was finally established that oxidation is a reaction in which electrons are given up.

When iron, for example, oxidizes, the iron atoms each give up two (and sometimes even three) electrons and become positively charged iron ions. The oxygen takes up these electrons and forms negatively charged oxygen ions (these are called oxide ions). The iron oxide formed during oxidation is held together by the opposite charges of the iron and oxide ions. It works somewhat more complexly, but fundamentally the same, when carbon is oxidized by oxygen. The carbon atom gives up a total of four electrons. However, these do not completely transfer to the oxygen atoms, so no charged ions are formed. Simply put, the carbon's electrons are now located between the carbon and the oxygen atom.[5] In this way, one carbon atom and two oxygen atoms are each connected to form a carbon dioxide molecule. As mentioned, the same type of reaction works not only with oxygen. The more or less same reaction can also be carried out with the elements fluorine, chlorine, or bromine, for example. All these substances can act as so-called oxidizing agents. An oxidizing agent is a substance that takes up the electrons given off by the substance being oxidized. The oxidizing agent itself is not oxidized but experiences the exact opposite: it is reduced (or expressed as a noun: it undergoes a reduction). It takes up electrons (the term reduction or reduce will become important later).

To summarize: When a humanoid breathes, they inhale oxygen-rich air. The oxygen then oxidizes various organic compounds in the body. It takes up electrons from these. Energy is released in the process. Part of this energy is released in the form of heat (which is why we get warm when we move, because more oxidation occurs in the body). The rest of the energy is converted into another chemical form, which we do not need to concern ourselves with at this point, and is used for all sorts of functions (e.g., muscle contraction).

However, we have just seen that oxidation is not necessarily just a reaction with oxygen, but generally a reaction with a substance that takes up electrons. This allows us to start thinking about whether an organism could also breathe something other than oxygen-containing air. For an inhaled substance, it would be helpful if it were a gas. This actually narrows our selection quite a bit. We need a substance whose boiling point is lower than the ambient temperature in which

---

[4] This is also where the term oxidation comes from. It is derived from the word Oxygenium, which is the Latinized form of a word actually composed of ancient Greek elements meaning "acid-forming" and ultimately nothing other than the Latin designation for oxygen.

[5] This process is also referred to as atomic bonding or covalent bonding. The negative charge of the electrons between the positive atomic nuclei is attracted by these two, so the atoms are bonded together.

our organisms live.[6] A substance that meets this condition would be, for example, chlorine. Anyone who paid a bit of attention in chemistry or history class[7] knows that chlorine is quite toxic.

But why is chlorine so toxic? Quite simply: Because it is a rather strong oxidizing agent. Chlorine loves to take electrons from other substances. Unfortunately, it usually does this uncontrollably and causes quite a bit of damage. However, if extraterrestrial organisms were to develop on a planet with a chlorine-containing atmosphere, it is conceivable that they would develop a biochemistry in which these undesirable reactions do not occur (or, if they do occur, are quickly countered by the organism's biochemical countermeasures). If chlorine is such a good oxidizing agent, then one could imagine that it could be used specifically for the oxidation of organic compounds in an organism. Instead of carbon dioxide (CO_2), tetrachloromethane (CCl_4) would then be produced. Its boiling point is about 77 °C. It could not be exhaled as easily as we exhale carbon dioxide. But it is not said that an organism must exhale all the products of oxidation that it uses for energy provision. We humans do not do that either. While we exhale the carbon dioxide, we excrete most of the water that is also formed with urine. An extraterrestrial, chlorine-breathing creature could do the same with its tetrachloromethane.

Chlor itself tends to oxidize other things a bit too strongly. Therefore, living beings would indeed have a hard time in a highly chlorinated atmosphere—even if they had originated on a corresponding planet and were adapted to it. But it doesn't necessarily have to be chlorine. Its neighbor below it in the periodic table of elements is, for example, bromine. It has very similar chemical properties but is not quite as reactive. Pure bromine would be liquid at our atmospheric pressure and room temperature. However, on another planets, a slightly lower pressure and a slightly higher temperature could prevail, and the bromine would evaporate. Moreover, it would not necessarily need to be pure bromine in the atmosphere.[8] Together with another, non-reactive gas, a certain proportion of bromine could thus remain in the atmosphere, even though the boiling point is below.[9] A certain dilution of bromine vapors by another gas could also be helpful because the still quite reactive bromine would thus cause less damage to the cells of living beings.

So we can conclude: Extraterrestrial organisms would not necessarily have to breathe an oxygen-containing atmosphere. There would be a number of other

---

[6] Ideally, the boiling point should be significantly below the ambient temperature. It would be very unpleasant if the atmosphere started to condense just because there was a slightly colder day on the planet.

[7] Chlorine was one of the many poison gases used in World War I to kill large numbers of people.

[8] The Earth's atmosphere, which we breathe, also consists of just a little over one-fifth oxygen. That is not only sufficient. It is even good because pure oxygen would again be too strong an oxidizing agent and would cause damage in the body.

[9] Just think of the evaporation of water. At 20 °C, water evaporates, although its boiling point is much higher, because the water vapor is diluted by the air, which simulates a lower pressure for the water.

2.1 Breathing Hydrogen

gases that could chemically be considered as alternatives, even if they would be deadly for life forms that originated on Earth.

What about the hydrogen that the Lothra breathe? Could there not be organisms somewhere in the universe that inhale hydrogen and use it to oxidize organic compounds? The answer is a clear no. Because hydrogen is not suitable as an oxidizing agent. Hydrogen is rather a reducing agent. We remember: Reduction was the reverse of oxidation. When a substance is reduced, it gains electrons.[10] Hydrogen can provide electrons for this. What hydrogen molecules, on the other hand, cannot easily do is accept electrons. Breathing hydrogen to oxidize organic (or even inorganic) substances simply does not work. A hydrogen-containing atmosphere is, after all, not an oxidizing but a reducing atmosphere.

So is Star Trek telling us nonsense at this point? Would the Lothra actually suffocate miserably? After all, one can inhale hydrogen without any immediate damage. But hydrogen cannot support the biochemical process of oxidation, which is the purpose of respiration. However, Jadzia Dax never said what the biochemistry of the Lothra actually does with the hydrogen. Realistically, the function of respiration can only lie in the energy supply of the body. Could this not also be achieved through reduction?

In principle, one might initially assume that the chances for this are not good. After all, reduction is the reverse of oxidation. The first law of thermodynamics tells us that energy is conserved. This means that the amount of energy released (usually as heat) in a reaction corresponds to the difference in energy content between the reactants and the products. Simplified, one can imagine it like this: If the reactants have an energy content of 100 kilojoules and the products only 70 kilojoules, then 30 kilojoules of heat must be released.[11] The analogous applies to the reverse path. If the reverse reaction is carried out to obtain the reactants again, then in our case, 30 kilojoules of energy must be supplied.

Biochemistry uses oxidation reactions for energy provision because they usually release a lot of energy. Conversely, this means that the same amount of energy would have to be put back into the reaction during their reversal (reduction). For a living being that wants to gain energy, this would be a highly disadvantageous approach. At least if it were to reverse the oxidation reactions that our

---

[10] At first glance, it may seem paradoxical that a process in which something is gained is called reduction. After all, to reduce means to decrease, not to increase. It becomes somewhat more understandable when one thinks of the history of the term. Reduction was originally understood (analogous to oxidation) as the removal of oxygen. In doing so, the mass of the reduced substance actually decreases.

[11] This notion is somewhat simplified because, strictly speaking, one cannot say that a substance contains this specific amount of energy. Energy has no zero point. Therefore, only energy differences can be specified. In practice, one helps oneself by indicating the energy of a chemical compound as the difference to the energy of the elements from which it is composed.

biochemistry uses for energy production. But elemental hydrogen does not appear in these reactions at all.[12]

What if our extraterrestrial organism did not simply release oxygen as elemental oxygen, but in combination with hydrogen? As a product of oxygen and hydrogen, water would be formed. In this case, we would couple the reduction reaction (which requires energy input) with the oxidation of hydrogen to water. It is well known that a lot of energy is released during the reaction of hydrogen with oxygen. In other words: hydrogen burns very well.[13] If this oxidation of hydrogen is now coupled with the reduction of an oxygen-containing organic compound, the energy requirement of the reduction can not only be compensated but even overcompensated. Since free, elemental hydrogen never appears in this reaction, there is no risk of explosion.

A hydrogen-breathing organism could use secondary hydroxyl groups as food, for example (this is a combination of an oxygen atom and a single hydrogen atom attached to the side of a chain of carbon atoms). Such secondary hydroxyl groups are abundant in all sugars. So, the Lothra could eat the same things we do, even if they breathe something different. The hydroxyl group could be cleaved off with hydrogen. What remains is a hydrocarbon, and water would be formed. Per mole[14] of hydroxyl groups, almost 90 kilojoules of energy would be released in the form of heat. The Lothra could use this energy.

Even though the formation of water usually releases a lot of energy, such reactions have a disadvantage. The oxygen must not be present as an elemental gas but must always be bound to an organic molecule (for example, in the form of a hydroxyl group). Extracting the oxygen from there requires a considerable portion of the energy that the hydrogen-breathing organism actually wants to use. Is there not an alternative that does not involve oxygen? What if the hydrogen did

---

[12] In our biochemistry, hydrogen indeed plays a very large role. However, it does not do so as elemental hydrogen, that is, as an H_2 molecule consisting of only two hydrogen atoms. Rather, hydrogen is part of larger molecules that, in addition to hydrogen, are primarily composed of carbon and usually oxygen.

[13] A popular example of this is the well-known image of the burning airship *Hindenburg*. The Hindenburg was filled with hydrogen instead of helium when it crashed in 1937. The photo of this disaster is very famous and still shapes the perception of hydrogen today. In fact, the problem was not the hydrogen itself, but the coating of the balloon fabric, which was supposed to prevent the hydrogen from slowly escaping. This coating could be ignited by static electricity and was also highly flammable. When the Nazi Propaganda Ministry had to explain the disaster, they decided not to publicly blame German engineering errors. Instead, they preferred to shift the responsibility to the American trade embargo, which prevented helium from being available in Germany, forcing the use of hydrogen. This decision still affects the public perception of hydrogen today.

[14] Since atoms and molecules are very small, individual atoms or molecules are not counted in chemistry; instead, $6022 \cdot 10^{23}$ are grouped together into one mole (this is a number with 24 digits). However, this is still not a significant amount of mass. One mole of water weighs just 18 g or, expressed as a volume, 18 milliliters.

2.1 Breathing Hydrogen

not extract oxygen from the molecules of the food but instead bonded with them directly?

Sugars and other carbohydrates are not suitable as starting materials for such a reaction, but there is a candidate in our food: fats. Not all fats, but at least the so-called unsaturated fats. Fat molecules consist of two parts. The base is a glycerol molecule. Three fatty acids are bound to this. There are many types of these fatty acids. However, the basic structure is always the same. At one end of the fatty acid is a so-called carboxyl group. This makes the fatty acid an acid and forms the connection to the glycerol. The rest of the fatty acid is a chain of carbon atoms. Often, 14, 16, or 18 carbon atoms are lined up in a row. However, fatty acids differ not only in the length of the chain (i.e., the number of carbon atoms). Strictly speaking, the chain does not consist only of carbon atoms linked together but of carbon atoms each with two hydrogen atoms attached (or three hydrogen atoms at the last carbon atom of the chain). This is the structure of a saturated fatty acid. In addition to saturated fatty acids, there are unsaturated fatty acids. In these unsaturated fatty acids, there are pairs of two adjacent carbon atoms, each with only one hydrogen atom attached. Simplified, one could say that they are missing two hydrogen atoms (which corresponds exactly to one hydrogen molecule).[15] An unsaturated fatty acid can now react with hydrogen and thereby become a saturated fatty acid.[16] This reaction is also energetically advantageous, so that about 125 kilojoules of heat are released per mole of hydrogen that reacts with an unsaturated fatty acid. This energy could also be used by a hydrogen-breathing organism to live on.

A few more details

The heat released during a reaction is equivalent to the reaction enthalpy. However, reaction enthalpy alone does not determine whether and how a reaction proceeds. There is another factor involved, which is called entropy.

Entropy is a somewhat difficult to understand but very important concept from a discipline called thermodynamics. Entropy is sometimes referred to as a measure of disorder. Let's simply imagine hydrogen. It is a gas, and in a gas, the individual molecules can move more or less freely. This is a very chaotic state, and the entropy of a gas is correspondingly very high. A fatty acid is, depending on the type of fatty acid, solid or liquid. In a liquid, the molecules can move significantly less freely. The entropy is correspondingly lower. In a solid, finally, the molecules are largely fixed to a specific location in the solid. They can, depending on the temperature, oscillate a bit around this location, but

---

[15] For this, the bond between the two carbon atoms is virtually doubled. This is why it is also called a double bond.

[16] The principle has been known for a long time. It has been applied for over 150 years, and the product is called margarine. Margarine is nothing more than fat in which all unsaturated fatty acids have been converted to saturated ones. This makes the fat more durable.

by and large, they do not have much room to move. A solid is therefore very ordered on a molecular level, and its entropy is low.

An important natural law, the second law of thermodynamics, states that entropy cannot decrease. When hydrogen reacts with an unsaturated fatty acid to form a saturated fatty acid, the entropy initially decreases. From a gas-phase molecule with high entropy and a liquid-phase molecule, a liquid-phase molecule is formed. Although the entropy of the resulting liquid-phase molecule (the saturated fatty acid) is somewhat higher than that of the original liquid-phase molecule (the unsaturated fatty acid), this small increase in entropy is not enough to compensate for the decrease in entropy due to the loss of the gas-phase molecule (the hydrogen). On balance, the entropy therefore decreases. So why can the hydrogen uptake still occur?

The heat released increases the entropy in the surroundings because the molecules move faster and thus more chaotically. This compensates for the decrease in entropy due to the hydrogen uptake and thus enables the reaction.

Biochemistry does not use reaction enthalpy alone but rather a reaction enthalpy corrected for the entropy effect, the so-called free reaction enthalpy (the free enthalpy G combines the information about the enthalpy H and the entropy S: $G = H - T \cdot S$, where T denotes the temperature). In the case of the saturation of a fatty acid by hydrogen, only about 85 kilojoules of usable energy remain from the 125 kilojoules of reaction enthalpy. ◄

Unfortunately, we cannot say exactly what kind of reaction the hydrogen-breathing Lothra use to gain energy. Jadzia Dax's account does not go into enough detail for that. What we can note, however, is that it is indeed possible to breathe hydrogen and live off it. However, not for us humans and most other humanoids who breathe oxygen in Star Trek. A hydrogen-breathing life form would need a completely different biochemistry than we know. Chemically, it would be possible, and who knows what kinds of different life forms have developed in the vast expanses of space.

Excursus

Can a planet even hold a hydrogen atmosphere?

An interesting question arises in connection with hydrogen-breathing life forms. Can the planet on which they develop hold its hydrogen atmosphere? The Earth's atmosphere is attracted by the Earth's gravity. Mars also had a proper atmosphere a long time ago. However, it has now become very thin. The problem is that Mars is smaller, and the gravity it exerts is therefore weaker. As a result, Mars's atmosphere has dissipated into space over time (a fate that will sooner or later also befall Melora's planet, the home of the Elaysians).

If the atoms or molecules of a gas have a high mass, they are attracted much more strongly. This is the reason why there is almost no helium in the Earth's atmosphere. Helium is actually the second most common element in the universe and is constantly being newly formed in the Earth's interior through radioactive decay processes.[17] The small, very light helium atoms, however, quickly escape into space, leaving behind the much heavier gases like oxygen and nitrogen. No wonder, since helium is the second lightest element of all. There is only one element that is even lighter: hydrogen.

The reason there is no elemental hydrogen in the Earth's atmosphere is not simply that it is so light. In our oxygen-rich atmosphere, it has largely converted into water. So hydrogen does indeed occur in the Earth's atmosphere, but not as elemental hydrogen, rather in bound form as water vapor. Since a water molecule is about nine times heavier than a hydrogen molecule, it does not escape into space as quickly.

In the atmosphere of the Lothra's homeworld, there is obviously no elemental oxygen, but there is elemental hydrogen. Wouldn't this hydrogen escape into space and leave the Lothra's planet to the same fate as Mars? Not necessarily. Because under one condition, a planet could also hold a hydrogen atmosphere: if its gravity were much stronger than that of Earth. It can therefore be assumed that the Lothra's homeworld is a very large planet. ◄

2.2 The Bussard Collector or Collecting from the Vacuum

The starships of Starfleet feature an extremely elegant design. The front, upper part forms a sometimes almost circular, sometimes slightly elongated disc, where a large part of life on board takes place. At its rear end, there is a connection to a second part of the spaceship. This slightly lower part is the drive section. Attached to it, on corresponding arms, are two elongated warp nacelles oriented in the direction of flight. This is the layout according to which almost all Starfleet ships are built. Let's take a closer look and examine a lesser-known detail on the warp nacelles.

As the name suggests, the warp nacelles are an essential component of the warp drive, which allows the spaceship to travel through space at faster-than-light speeds. Most of the warp nacelles on 24th-century starships are occupied by a bluish glowing part, which apparently has to do with the warp drive itself (at least it lights up when the ship goes to warp and shines in the same blue tone as the warp core of the Enterprise-D). The front part of the warp nacelles usually receives far less attention. On many ships like the Enterprise-D or the Voyager, the front part

---

[17] When heavy, radioactive elements decay by alpha decay, an alpha particle is released, which is nothing more than the nucleus of a helium atom. As soon as it has collected two electrons from the surroundings, a normal helium atom is formed. Since such alpha decays occur in large numbers in the Earth's interior (and have occurred throughout Earth's history), a considerable amount of helium has accumulated in natural gas. Depending on the natural gas deposit, helium can make up to 16% of the natural gas. Technically, helium is obtained by separation from natural gas.

does not glow blue but red. But what is actually behind the red glowing tip of the warp nacelles?

One might suspect that it is used for flying backward. This assumption would at least be close, as the impulse drive on many ships glows similarly red, and a forward-facing thrust drive would indeed make the ship fly backward. However, upon closer inspection, one finds something else here. The so-called Bussard collectors.

These are named after the American scientist Robert W. Bussard, who passed away in 2007. Although he did not develop the Bussard collector (Star Trek engineers will invent that in the future), he was involved in the research of nuclear fusion. Shortly before Gene Roddenberry had Captain Christopher Pike travel to the planet Talos IV in the first pilot film *"The Cage"*[18], Robert Bussard proposed a novel propulsion concept for spaceships in 1960. In this concept, interstellar hydrogen was to be used in a fusion reactor to propel the spaceship. For this, the interstellar hydrogen first needs to be collected. This is what the Bussard collectors are for.

The operating principle of the warp drive then differs significantly from Bussard's concept. First, hydrogen is not used for energy provision through nuclear fusion but through an antimatter reaction. Apart from a considerable risk, this is a quite clever approach by Starfleet engineers. While a lot of energy is released during nuclear fusion, much more energy is obtained through the complete annihilation by antimatter. Furthermore, the warp drive is not based on a ramjet engine as proposed by Bussard. He could not have known what Zefram Cochrane would know over a hundred years later when inventing the warp drive.

The basic idea is definitely good. The propulsion of a spaceship and the many other ship systems require a lot of energy, and the next space gas station is usually quite far away. It makes sense to try to use what is found in space. The most common chemical element in the universe by far is hydrogen. The second most common chemical element is helium. However, it is significantly rarer and not really suitable for power production.

The hydrogen is not to be collected from suns or on any planets. The Bussard collectors are about interstellar hydrogen. In other words: hydrogen in the space between the stars. But isn't space actually a vacuum?

That's correct. It is even an incredibly good vacuum. When vacuum technicians on Earth today try to create an ultra-high vacuum, they have to go to great lengths. And in the end, they still have a few residual molecules in the system.[19] Compared to that, the interstellar vacuum is a vacuum of the highest quality. So is it worth

---

[18] *"The Cage"* is sometimes listed as the 1st episode, sometimes as the 0th episode of the 1st TOS season.

[19] Ultra-high vacuum plays a role in chemistry, for example, in the study of surface structures of catalysts. If there were still air between the sensor and the atoms of the structure being studied, it would significantly distort the result. On the one hand, one would partially measure the molecules of the air instead of the atoms of the surface, and on the other hand, electron beams used for the study would collide with the air molecules and be deflected.

2.2 The Bussard Collector or Collecting from the Vacuum

collecting the remaining gas from this vacuum? After all, there is almost nothing there.

The argument is correct on one side but overlooks the incredible size of space. If the spaceship only collects the hydrogen that happens to fly into the Bussard collector, it is not worth it. However, the Bussard collector collects hydrogen from a wide area using magnetic fields. Even if one considers that there is almost nothing per cubic meter of space, space still has an unimaginable number of cubic meters. Therefore, interstellar hydrogen provides an indescribable reservoir of a valuable energy carrier. This should, of course, not be left unused.

Nevertheless, the amount of hydrogen per cubic meter is actually very small. This leads to some problems. To understand these, we first need to take a look at chemical thermodynamics. Chemical thermodynamics is a discipline of chemistry that deals, among other things, with the question of whether a reaction can occur at all or what happens when a mixture starts to evaporate.[20] Chemical thermodynamics ultimately always asks whether molecules move from state A to state B under certain conditions. State A in our case is hydrogen, distributed in the infinite expanses of space. State B is hydrogen molecules in a hydrogen tank on board a Starfleet spaceship. Let's first take a look at state A.

In interstellar space, there are on average one million particles per cubic meter[21]. That doesn't sound too bad. A million is quite a lot, after all. Moreover, matter in space is not evenly distributed. There are indeed clusters of particles where the particle density is significantly higher. It becomes very high, of course, in stars or planets. But even away from classical celestial bodies, there are large areas with increased particle density. Just think of the so-called nebulae. The most famous example, the Andromeda Nebula, is a bad example for this. Because it is a galaxy, a collection of stars that only looks a bit nebulous from Earth. However, within the galaxy itself, there are a large number of "real" nebulae; these are areas with increased particle density without being a star or a collection of stars. These nebulae could have formed, for example, as remnants of a supernova or as a precursor to the birth of a star that has not yet formed. The chemical composition of the nebulae can differ somewhat. If it was formed by a supernova, there will be more heavy elements like carbon, oxygen, or iron in the corresponding nebula. The main component, however, is still the lightest of all elements: hydrogen. In cosmic nebulae, we are not just dealing with a million particles per cubic meter.

---

[20] Chemical thermodynamics explains, for example, why in the evaporation of a mixture of water and alcohol, the alcohol does not evaporate first and then the water. Even if you are below the boiling point of water but above that of alcohol, both will always evaporate. The alcohol merely accumulates somewhat more in the vapor phase than the water. Conversely, when cooking a wine sauce, the alcohol does not eventually "boil off." What you taste in the sauce is not some ominous alcohol flavor that would remain after the alcohol evaporates, but simply the alcohol that has not completely evaporated. The idea that there would be a clean separation is a widespread but completely nonsensical misunderstanding.

[21] Particles here do not necessarily mean particles but can refer to individual atoms, molecules, or ions, i.e., electrically charged atoms or molecules.

Here we are talking more about 100 million particles minimum. The number of molecules can rise to a trillion particles. A million particles in "normal" interstellar space, a billion or even a trillion particles within nebulae: that's something to work with.

Really? What does it mean if there are a million molecules in a cubic meter? We should not forget that atoms and molecules are very, very small. Let's convert this into amounts of substance. Because chemists do not want to calculate with individual molecules and atoms, as the numbers would simply be too large, they express particle numbers as amounts of substance. This amount of substance has a unit called *mol*. Behind the mole is—simply put—just a very large number. Since 2019, the mole has been directly defined as a number: 6.02214076 times ten to the power of 23 particles make up a mole. Scientists always use this notation when they want to express very large (or small) numbers. Simply put, it is approximately a 6 with 23 zeros behind it (in words: 602 sextillion). That's quite a lot. At least as a number. But how much matter is that actually? Until 2019, the mole was defined differently. The amount was ultimately the same, though. Back then, the mole was defined as the number of atoms in twelve grams of carbon.[22] 602 sextillion carbon atoms thus correspond to exactly twelve grams. The hydrogen atom is significantly lighter than the carbon atom: 602 sextillion hydrogen atoms together make up only about one gram, while 602 sextillion hydrogen molecules at least weigh two grams, as they consist of two hydrogen atoms. Suddenly, a million particles in a cubic meter doesn't seem like much anymore.

Considering that a sextillion is a quadrillion times a million, it becomes clear how much space you have to harvest to get a usable amount of hydrogen. If you want to collect just one gram of hydrogen from interstellar space, you have to graze several hundred quadrillion cubic meters. Even in a very dense nebula, you would still have to collect the gas molecules from several hundred billion cubic meters. An absurdly large volume? There can't be much hydrogen in interstellar space, can there? One is amazed at how much there actually is.

One simply has to consider how many cubic meters are available in space. The star closest to our solar system is Proxima Centauri. Between us and this star lies about 4.2 light-years. That is a proud 40 trillion kilometers or 40 quadrillion meters. If one imagines a cube between our solar system and Proxima Centauri with this side length, it would have a volume of about 75 cubic light-years. That doesn't sound like much at first. However, if you convert it into cubic meters, you get a number that most people have probably never heard of: about 65 octillion cubic meters. That is a six with an impressive 49 zeros behind it. The ridiculous hundred quadrillion cubic meters for one gram of hydrogen can be found several hundred quintillion times in this cube. Even if every person on earth had their own

---

[22] This carbon had to be isotopically pure. That means that only the most common carbon isotope C-12 was allowed to be present in this gram.

2.2 The Bussard Collector or Collecting from the Vacuum

spaceship and flew to Proxima Centauri, each collecting one kilogram of hydrogen (with which one could provide an indescribable amount of energy in a fusion reactor), the interstellar hydrogen in this cube would still be enough for many hundreds of billions of round trips for each individual person.

The amount of hydrogen in interstellar space is therefore easily sufficient for more spaceships than we can imagine. So where is the challenge?

The net movement of many particles always follows a gradient. The term gradient refers to the change of a quantity over a distance in space. The easiest way to illustrate this is with the example of heat flow (even though no particles flow in this case): If an object is hot at one end and cold at the other, there is a temperature gradient within it. Following this temperature gradient, heat flows from the end with the high temperature to the one with the low temperature. The principle can be transferred to other quantities. In chemistry, for example, the concentration gradient is very important. Let's imagine a cup of tea (*Earl Grey, hot;* just as Captain Picard loves it). We add a sugar cube to it. In the tea around the sugar cube, there is quickly a high sugar concentration, while tea that is further away is still practically unsweetened. The sugar molecules now move along their concentration gradient away from the sugar cube to the less sweetened part of the tea. The molecules thus migrate from an area of high concentration to one of low concentration. The opposite effect can never be observed, just as one cannot observe heat voluntarily flowing from an area of low temperature to an area of high temperature against the temperature gradient. We call this movement of molecules diffusion, and it follows the concentration gradient.[23] For our question about collecting interstellar hydrogen, another gradient is crucial: the pressure gradient. Flows always occur along pressure gradients. Simply put: A gas always flows from an area of high pressure to an area of low pressure. This is quite comparable to diffusion along the concentration gradient. In a gas, there are more molecules in the same volume at high pressure than at low pressure. By increasing the pressure (a bit simplified), the concentration increases. The molecules therefore move analogously from the area of high pressure to an area of low pressure. This is exactly where our problem lies.

---

[23] Strictly speaking, diffusion actually follows the gradient of the chemical potential, not the concentration gradient. Since the chemical potential of a substance is, simply put, high when the concentration of the respective substance is high, it is almost always sufficient in practice to assume that diffusion follows the concentration gradient. This no longer works when we think of extraction, for example. In this case, molecules diffuse from a solvent with low solubility into a solvent with higher solubility. At the so-called phase boundary between the two immiscible liquids, the molecules may then diffuse from an area of low concentration to an area of high concentration. This is because the chemical potential does not depend solely on the concentration but also on which other (solvent) molecules surround it. In a solvent with high solubility, the chemical potential is, simply put, lower at the same concentration than in one with low solubility. Apart from this special case, the assumption that diffusion follows the concentration gradient is usually a practical simplification.

Let's think about the opposite of collecting interstellar gas into the spaceship. So let's imagine the escape of gas from the spaceship. This is exactly what happens when an airlock is opened. This is very impressively shown in the 23rd episode of the 2nd season of ENT, *"Regeneration."* There, two Borg drones are blown into space after Malcolm Reed opens an airlock. There is a strong airflow along a pressure gradient. From high pressure on board to the very low pressure of space.

If you want to collect interstellar gas, you are always fighting against the pressure gradient. You have to try to transport the gas from the very low pressure of space into the inevitably much higher pressure of the storage tank (if the pressure in the storage tank were not much higher than in space, there would again be just as few molecules per cubic meter in it, making it practically empty and therefore useless). The interstellar hydrogen must therefore flow against the pressure gradient.

To make a gas flow from an area of low pressure to an area of high pressure is indeed possible. Just think of a compressor. When a compressor fills a compressed air bottle, it sucks in air at atmospheric pressure and presses it into a steel bottle, where the pressure inside may be more than a hundred times higher. A flow from low pressure to high pressure is certainly possible. It merely requires a certain amount of effort in the form of energy and appropriate apparatus. This effort becomes greater the more the pressure needs to be increased. Now, someone might argue that the pressure in the case of Bussard collectors does not need to be increased significantly at all. Let's assume that the storage tank to be filled has a pressure of 1 bar (which corresponds roughly to the pressure of the Earth's atmosphere). The interstellar gas has a pressure of nearly 0 bar. Thus, there is a pressure difference of about 1 bar. The compressor, which has to fill the compressed air bottle with, for example, 200 bar, on the other hand, has to overcome a pressure difference of 199 bar. So, one might think that the compressed air compressor has to exert much greater effort.

In fact, the effort to transfer a gas from a low pressure to a high pressure is not dependent on the pressure difference but on the pressure ratio. The air compressor has to increase the pressure of the air two hundredfold. That is quite something, but the compression from the interstellar vacuum can only laugh at that. So, the question is: What is the actual pressure of interstellar hydrogen?

The pressure depends on two factors. One is the number of molecules per volume. As we have already seen above, this is very low. The second factor is the temperature. Klingon General Chang hints at the temperature in space in the sixth Star Trek movie *"Star Trek VI: The Undiscovered Country"* when he says: "In space, all warriors are cold warriors." Even if he actually wants to express something else, he ultimately expresses very well that it is very cold in space. The temperature of the cosmic background radiation, which somehow represents the temperature of everything that exists between the stars, is about 2.7 K: Less than minus 270 degrees Celsius and thus just slightly above absolute zero. If you put this temperature together with the number of particles per volume into the equation

2.2 The Bussard Collector or Collecting from the Vacuum

of state of the ideal gas, you get a pressure of less than one zeptobar.[24] Written out, we are talking about a pressure of about 0.000000000000000000004 bar. Even in a cosmic nebula, there is only a pressure of less than one femtobar. That is one millionth of a nanobar. The compression of this interstellar gas to atmospheric pressure represents only a comparatively small change in pressure when viewed as a pressure difference. For the pressure ratio that is decisive for the energy requirement, however, the difference is enormous. Accordingly, the energy requirement is high.

However, the effort is not only expressed in a high energy requirement. We are also dealing with a tremendous technical effort. How is the compressor of a Bussard collector supposed to work at all? The classic mechanical compressor types are out of the question. Just imagine a screw compressor trying to convey a gas between two screws when the average distance between the molecules is greater than the distance between the two screws. This problem affects all mechanical compressors in principle. Here, only chemistry can actually help. One conceivable method could be a so-called electrochemical compressor. In this process, hydrogen is transported through a membrane, as known from fuel cells. Unlike in a fuel cell, however, the electrochemical compressor does not release electrical energy but absorbs energy. With the help of this energy, it transports the hydrogen through the membrane from an area of low pressure to an area of high pressure. The energy requirement would still be considerable with our pressure ratio, but at least it might be possible to carry out the compression electrochemically.

However, due to the low density that hydrogen exhibits in space, it is not enough to simply mount an electrochemical compressor at the tip of the warp nacelle and fill its hydrogen tank with it. If you were to collect only the hydrogen that is exactly in the flight path of the spaceship (or even only in the much smaller path of the Bussard collector), you would collect almost nothing. The engineers of Starfleet have, of course, thought of this and therefore collect the interstellar gas with enormous magnetic fields.

If you take a closer look at the hydrogen in space, you will find that it largely does not consist of uncharged molecules of two hydrogen atoms. Rather, a significant portion is ionized by cosmic radiation. So, we are not dealing with uncharged atoms or molecules, but with electrically charged ions. These can indeed be influenced by a strong magnetic field. If this magnetic field is not constant (which would be the case just by flying the spaceship), it would also set the hydrogen ions in motion. The cause of this is the so-called Lorentz force (named after the Dutch physicist Hendrik Antoon Lorentz). If the whole thing is cleverly designed, one could well imagine that interstellar gases are directed to the Bussard collectors, where they are transported into a storage tank by an electrochemical compressor. However, the spaceship should not be moving at warp speed, as it would otherwise rush past the interstellar hydrogen at faster-than-light speed. However, Robert Bussard did not think of faster-than-light speed in his concept anyway.

---

[24] The prefix *zepto* stands for one sextillionth.

> **Excursus**
>
> **How is hydrogen stored at all?**
> Once the interstellar hydrogen has finally been collected (or simply refueled at the last visit to the starbase), it needs to be stored somewhere. How is that actually done?
>
> Hydrogen storage is a complicated matter. There are a whole range of methods to do this. The most common method today is to store hydrogen gas at high pressure in a pressure-resistant and tight tank. However, even at a pressure of 300 or even 700 bar, as found in the tanks of modern hydrogen cars, there is still comparatively little hydrogen in the tank. The density of hydrogen is simply too low. An alternative is cryogenic hydrogen. In this case, the hydrogen is cooled to about minus 253 degrees Celsius and thus liquefied. This allows for slightly more hydrogen in the same tank volume compared to the pressure variant. However, the energy requirement is significantly higher and there are storage losses because gas must be released repeatedly as liquid hydrogen continuously evaporates.
>
> To address the problem of the low density of hydrogen in conventional storage, chemistry has developed a number of methods. Simply put, hydrogen is chemically bound to a carrier. This can be a metal, for example. The very small hydrogen molecules settle into the spaces between the metal atoms and can be stored there in much higher numbers per volume. To release the hydrogen, the so-called metal hydride only needs to be heated. Another possibility is so-called Liquid Organic Hydrogen Carriers (LOHC, named after the English term "liquid organic hydrogen carrier"). In this case, hydrogen is bound to a liquid organic compound through a chemical reaction called hydrogenation. The carrier liquid is chemically transformed in the process. An aromatic compound becomes the corresponding saturated compound. In the simplest (though technically not sensible, because carcinogenic) case, benzene would be converted to cyclohexane. In practice, substances like dibenzyl toluene are used as LOHC. These LOHCs can be hydrogenated and thus absorb a lot of hydrogen. One molecule of dibenzyl toluene can, for example, absorb up to nine molecules of hydrogen. The hydrogenated, hydrogen-rich form of the LOHC can then be stored under ambient conditions without high pressure or low temperature. As a liquid, the LOHC is also easy to handle and can be transported relatively easily. To retrieve the hydrogen, the hydrogen-rich LOHC must be heated and the reverse reaction of hydrogenation, called dehydrogenation, must be carried out with the help of a catalyst. The hydrogen can then be used energetically, for example by feeding it into a fuel cell. The dehydrogenated LOHC can be stored and reloaded with hydrogen through hydrogenation when needed.
>
> Such hydrogen technologies are enormously important for establishing a functional infrastructure for a future energy system in which hydrogen is to play a major role. But does a spaceship that wants to operate a fusion reactor with hydrogen need such hydrogen storage? As a scientist who works on

chemical hydrogen storage himself, I say this very reluctantly, but the answer is a clear: No!

Why? The hydrogen for a fusion reactor could simply be bound to oxygen and stored as water. With electrolysis, hydrogen can be easily obtained from water when needed. This would make no sense at all in our current energy technology (and the energy technology of the next decades). Electrolysis requires significantly more energy than a fuel cell delivers. Even if electrolysis and fuel cell technology are pushed to the limit, chemical thermodynamics shows that the maximum energy that can be obtained from the fuel cell is what was put into the electrolysis. A fusion reactor, on the other hand, generates such a large amount of heat from hydrogen that the energy requirement of electrolysis does not matter at all. Since water is the easiest to store, a fusion reactor would probably not use any overly complex storage technology. ◄

2.3 Explosions in Space

As we have seen, interstellar hydrogen is often collected in Star Trek. Apparently, other things can be done with the hydrogen distributed in space. In the 5th episode of the 2nd DSC season, *"Old Friends,"* the Discovery pursues a shuttle in which they suspect Mr. Spock. He had been wrongly accused of several murders shortly before, which is why he was on the run. However, at the time the shuttle was intercepted by the Discovery, Mr. Spock was not in it. In fact, the shuttle was being piloted by Philippa Georgiou, the deposed Terran Empress who now works for a Starfleet intelligence service. When she does not respond to calls, Captain Pike decides to stop the shuttle with a photon torpedo detonating nearby. The plan succeeds, but a Terran Empress does not give up so easily. To escape, she ignites the hydrogen of the nebula she had just flown into.

Again, a whole series of questions arise. Why didn't the photon torpedo ignite the nebula (if it is even possible to ignite it)? Even more interesting is the question of how it is possible to ignite the hydrogen of an interstellar nebula at all.

The explosive capability of hydrogen is generally well known. The problem with igniting interstellar hydrogen should also be clear to some: The explosion of hydrogen is a so-called oxyhydrogen explosion. This involves a sudden combustion in a mixture of hydrogen and oxygen. The problem with igniting interstellar hydrogen is obviously the lack of oxygen. On Earth, this is usually not such a big problem. After all, air consists of about 21% oxygen. In space, however, this element is very rare. Hydrogen is known to be the most common element in the universe. Helium follows in second place. Then there is a long gap. When comparing the total amount of individual elements in the universe, these two elements make up almost the entire universe. Oxygen is just a footnote in the composition of the cosmos.

Now, one might argue that there are also larger amounts of oxygen on Earth. However, oxygen (and other elements) are very unevenly distributed in the universe. If the Earth's atmosphere is an area of the universe with an increased concentration of oxygen, then it can also exist in other areas of the universe, can it not? One should not overlook the mass of the elements. Hydrogen is the lightest element, helium the second lightest. An oxygen atom has 16 times the mass of a hydrogen atom and still four times the mass of a helium atom. Consequently, oxygen atoms are attracted much more strongly in gravitational fields than is the case with hydrogen or helium. That is why there is hardly any helium in the Earth's atmosphere, even though this element is so common in the universe. Because it is so light, the atmosphere loses it to space. Oxygen, on the other hand, is significantly heavier and can therefore accumulate in the Earth's atmosphere. What works on Earth can fundamentally happen just as well on other planets. Therefore, it is quite plausible that there are so many planets with an oxygen-nitrogen atmosphere in Star Trek. A cosmic nebula, on the other hand, is virtually an atmosphere without a planet. Due to the absence of a large, central mass, like a planet, there is no mechanism that would enrich oxygen compared to hydrogen in a cosmic nebula.[25] The hydrogen in the nebula is therefore practically only mixed with helium. There should hardly be any oxygen.

The fact that there are probably hardly any significant concentrations of elemental oxygen in a cosmic nebula, which consists largely of hydrogen, has another reason. That is simply stability. If it were possible to ignite the corresponding oxyhydrogen mixture, it would have happened long ago in the millions of years that the nebula has existed. And even if it did not occur in the form of an explosion, there would certainly be a slow reaction over millions of years. Water might not form in a large explosion. However, over a period of millions of years, hydrogen and oxygen (if it is possible for them to react in this nebula[26]) would indeed react. Even if it were only a creeping process. In the end, the nebula would then consist of water and not of hydrogen and oxygen.

The real problem with creating an explosion by igniting interstellar hydrogen is the density. Let's assume that the nebula contained enough oxygen in addition to hydrogen. Perhaps even a precisely tuned, stoichiometric mixture. Could this oxyhydrogen gas then really explode?

In space, as we have already seen, there is very low pressure and consequently the density is very small. This initially affects how much energy is released per

---

[25] Any astrophysicists among the readers may forgive the highly simplified picture of the "atmosphere without a planet." In fact, many nebulae have a kind of central star. For example, nebulae can arise from a supernova. At the center of the nebula, there is then a neutron star or a black hole. From there, a tremendous gravitational force emanates. However, a nebula that arose from a supernova has a very clear direction of movement away from the center. Unlike the atmosphere of a planet, which forms by "collecting from the outside," a cosmic nebula did not form by collecting heavy elements from the surrounding space. In fact, there can be quite a few heavier elements in nebulae that originated from supernovae. These are actually primarily formed in supernovae. However, an ignitable oxyhydrogen mixture should not form in the process.

[26] The chemical reaction equilibrium could prove to be a problem here. We will address this aspect shortly under the keyword "Le Chatelier's Principle."

2.3 Explosions in Space

volume. The amount of energy released in a chemical reaction is determined by two factors. One is the reaction heat per amount of hydrogen converted, and the other is the amount of hydrogen. Since the amount of hydrogen per volume in space is very low, the same applies to the amount of heat released per volume. Let's take the assumptions about the density of cosmic matter from the previous section again. Then, in an oxyhydrogen reaction, even at the highest density that occurs in cosmic nebulae, only about 40 microjoules per cubic meter would be released in the form of heat. That is an amount of energy needed (under Earth's gravity) to lift an object with a mass of four milligrams one meter high. In other words: almost nothing. With an explosion of such intensity, the question arises whether one would even notice it. How it is supposed to stop a spaceship like the Discovery in any way is then another question entirely.

However, we do not want to deal with the question of how much (or how little) explosion a Starfleet spaceship can withstand. We are concerned with questions of chemistry, and in this context, there is another question: Can the chemical reaction of the oxyhydrogen gas even take place?

The problem in this context is Le Chatelier's principle. This concept, named after the French chemist Henry Le Chatelier, states that the position of an equilibrium shifts in such a way that it avoids an external constraint. First of all: What is an equilibrium in chemistry?

Any reaction that can proceed forward can also proceed backward. The more product that has been formed, the more the reverse reaction begins to occur. At some point, the same amount of product is converted back into reactants per unit of time as reactants are converted into products in the same unit of time. The net reaction comes to a halt.[27] Consequently, a reaction can never proceed completely but only up to equilibrium. This equilibrium can, as in the case of the hydrogen-oxygen reaction, lie so far on the side of the products (i.e., water) that one might get the impression that there is no corresponding limitation.

However, the exact position of the equilibrium between no reaction and complete reaction is not fixed. Here, Le Chatelier's principle comes into play. The external constraint in our case is the very low pressure. When the pressure is lowered, a reaction system in equilibrium tries to compensate for this by increasing the number of gas molecules. Through the additional gas molecules, it essentially fights against the low pressure. In the hydrogen-oxygen reaction, two hydrogen molecules react with one oxygen molecule to form two water molecules. Three molecules become two. This is the exact opposite of what the system wants according to Le Chatelier at low pressure. It must therefore do the opposite. If

---

[27] Indeed, any reaction that can proceed forward can also proceed backward. Strictly speaking, however, this is not even necessary to have a reaction equilibrium. The equilibrium is established when the reaction reaches a minimum of Gibbs free energy. Such a minimum exists for every reaction and is always somewhere between no reaction and complete reaction (never exactly at no reaction or complete reaction). However, the corresponding excursion into chemical thermodynamics goes a bit too far here and would require some mathematics.

water is split into oxygen and hydrogen, then two molecules become three. This is the favored reaction at low pressure. Therefore, the equilibrium of the hydrogen-oxygen reaction begins to shift more and more to the side of the reactants (away from water, towards hydrogen and oxygen) as the pressure decreases. If the pressure is lowered slightly below atmospheric pressure, this is hardly noticeable. However, in space, even in a relatively dense cosmic nebula, there is an unimaginably low pressure. Therefore, the position of the reaction equilibrium shifts massively towards the reactants. The hydrogen-oxygen reaction can occur, but the maximum conversion allowed by the equilibrium is very low.

Some more details

Le Chatelier's principle not only says something about the influence of pressure on the equilibrium but also about the influence of temperature. The hydrogen-oxygen reaction is exothermic. This means that when it occurs, heat is released. In space, the temperature is very low. This low temperature is again an external constraint that the system tries to fight against by shifting the position of its equilibrium. Since heat is released during the forward reaction (water formation), it can fight against the low temperature. This circumstance shifts the equilibrium back to the product side at low temperatures. And since it is very cold in space, the equilibrium position shifts very strongly towards the product water due to the low temperature.

The low temperature could theoretically compensate for the low pressure. Superficially, this would make almost complete reaction possible again. However, at low temperatures, reactions also proceed more slowly. At about minus 270 degrees Celsius, as it is in space, one can hardly speak of an explosion in such a slow reaction. Moreover, the temperature would rise due to the reaction. The pressure, however, would hardly rise. Therefore, the equilibrium position could not effectively be shifted so much towards water that a real explosion would be possible. The limitation of the reaction yield thus remains. ◄

Another aspect must be considered when looking at the combustion of an interstellar nebula. The low pressure causes even more problems. Even if we imagine a nebula consisting of a flammable mixture of oxygen and hydrogen and assume that the reaction equilibrium is not a problem, another problem arises: the reaction rate.

No chemical reaction occurs instantaneously. Time always passes. This can be a lot of time or very little. Similar reactions can proceed at very different rates. This is referred to as the respective reaction kinetics. When a nail rusts, it is an oxidation of iron. The reaction takes months or years to complete. When hydrogen is oxidized, it is the hydrogen-oxygen reaction discussed in this section. This reaction proceeds—literally—explosively. Within a few milliseconds, the reaction of

2.3 Explosions in Space

oxygen with hydrogen is complete. The reaction of oxygen with iron, on the other hand, takes years. This has various reasons.

One significant point is accessibility. The outermost layer of iron atoms in the nail is still easily accessible to the oxygen in the air. With the next layer, it becomes more difficult, and the deeper it goes into the nail, the worse the accessibility becomes. Accordingly, the oxidation reaction in the case of the iron nail proceeds very slowly because the oxygen hardly reaches the iron atoms in the nail. In contrast, if you look at iron filings, you will find that they oxidize completely much faster. Due to the many small iron particles, there is a much larger total surface area, and the deepest iron atoms are also not as deep inside the filing as in the case of the nail. Accordingly, iron filings oxidize faster than an iron nail. The smaller the individual iron particles become, the larger the total surface area and the shorter the path to the interior. The smallest particles one can imagine are individual atoms. Or in the case of oxygen and hydrogen: individual molecules. If the molecules are perfectly mixed, accessibility is no longer a problem, and the reaction can theoretically proceed at a very high speed. However, there are other factors that limit the speed of a chemical reaction. One of them is concentration.

The relationship between concentration and reaction rate is probably not very surprising: The higher the concentration of the reactants, the faster the reaction proceeds. Conversely, the lower it is, the slower the reaction. The more molecules of the reactant are present per volume, the higher the probability that they will collide, allowing more reaction to occur per unit of time. This relationship is actually quite intuitive. But what does pressure have to do with it?

Simply put, pressure corresponds to concentration in gases. More precisely: the partial pressure. This is understood as the pressure multiplied by the proportion of the substance in the mixture. Why this partial pressure is essentially the measure of the concentration of a substance in a gas mixture can be understood from the example just given. Concentration (in liquids) ultimately indicates how many molecules are present per volume element (for example, per liter). Since the volume of a liquid hardly changes with an increase in pressure, the concentration also does not change when the pressure increases. It is different with a gas. Here, the volume does change when the pressure changes. If the pressure decreases, the volume increases. However, the total number of molecules remains unchanged with a change in pressure. Thus, the number of molecules per volume element decreases. Consequently, the probability that a molecule will collide with another and a chemical reaction can occur decreases. Therefore, partial pressure can be thought of as a kind of effective concentration in gases.

Now let's imagine our hypothetical oxyhydrogen cloud. This consists of a stoichiometric mixture of hydrogen and oxygen. That means we have two-thirds hydrogen and one-third oxygen. In terms of the mixing ratio, everything is such that the interstellar oxyhydrogen cloud could burn perfectly. However, the pressure is very low. And low means really low in the context of pressure in space. Accordingly, the partial pressures are very low. The partial pressure of hydrogen would be only two-thirds of the already tiny total pressure. The partial pressure of oxygen would be only one-third of the tiny total pressure. The distances between the hydrogen and oxygen atoms are enormous.[28] Accordingly, the molecules rarely collide. The reaction may be possible, but it is very slow. It would have little to do with a rapidly occurring reaction. The explosion of an interstellar cloud of hydrogen and oxygen would therefore probably not be a very impressive event. At least a large shock wave would not be expected. Rather, a slow burn that glows so faintly that it is hardly noticeable. After all, only a small part burns due to the reaction equilibrium. And even if everything were to burn, it would still be almost nothing in terms of energy per cubic meter. Even if it were possible to ignite the hydrogen in a cloud and the combustion proceeded quickly, the result would probably still not be a particularly impressive explosion.

> **Excursus**
>
> ### What else could explode?
> As we have seen, hydrogen clouds in space are not really suitable for ignition. But could there be cosmic clouds with a different composition that would be suitable for this?
>
> In the ninth feature film *"Star Trek: Insurrection"*, the Enterprise is pursued by two Son'a ships. To get the situation under control, William T. Riker develops a tactic referred to by Geordi La Forge as the Riker Maneuver. For this, the Enterprise collects interstellar metreon gas with its Bussard collectors. The metreon gas is then released directly in front of the Son'a's bow, which ignite it with their own weapons and cause it to explode. How can this work? And what exactly is metreon gas?
>
> In the Voyager episode *"Jetrel"*, we learn that unstable metreon isotopes can be used to create a weapon of mass destruction called the metreon cascade. This tells us a lot about what metreon is chemically. It is obviously not a compound made up of atoms of different elements. Metreon is apparently an element itself. Isotopes are different atoms that still belong to the same element. The various isotopes of an element have the same number of positively charged protons. As a result, they also have the same number of electrons, which in turn determine the chemical properties. Chemically, isotopes are atoms of the same element that differ only in their mass because they have different numbers of neutrons in the nucleus. The chemical properties, however, are the same.[29]
>
> Metreon thus seems to be a previously unknown chemical element. Like all elements that are newly discovered today and probably in the future, it is

---

[28] The term enormous is meant chemically. Of course, the average distance between two molecules in an interstellar cloud is only a few millionths of a meter. For a molecule that is only a few hundred billionths of a meter in size, however, this is an unimaginably large distance.

[29] That is not entirely true. There is the so-called kinetic isotope effect. Heavy isotopes generally react more slowly chemically than lighter isotopes of the same element. This effect can be noticeably observed at least in hydrogen isotopes.

2.3 Explosions in Space

unstable. However, the radioactive decays that transform the atoms of one element into atoms of another element cannot be triggered chemically. To trigger a metreon explosion, one would have to initiate nuclear fission.[30] If the Son'a's weapons release neutron radiation, this could indeed happen.

Such a "nuclear chemical" process, in which a "reaction" occurs at the nuclear level, would be a possible explanation. But could the metreon explosion also be triggered by a "classical chemical reaction"?

Just because metreon is an element does not mean that metreon cannot exist in the form of molecules. Many elements do this after all. In many nonmetals, two atoms of an element combine to form a molecule of the element. Just think of hydrogen (H_2), oxygen (O_2), nitrogen (N_2), or chlorine (Cl_2). For individual elements, it can even be significantly more atoms. For example, oxygen can also exist as ozone. An ozone molecule consists only of oxygen atoms. However, it is made up of three oxygen atoms (O_3). Sulfur, for example, forms ring-shaped molecules of eight sulfur atoms (S_8) under certain conditions. Metreon atoms may combine in a similar way to form molecules.

The formation of such metreon molecules would be associated with the formation of bonds between the atoms. When bonds are formed, heat is released. The reaction is exothermic, as they say in chemistry. If the formation reaction of metreon molecules is very strongly exothermic and proceeds very quickly, then one could at least imagine an explosion. However, in the negligibly small pressure of space, the reaction equilibrium would again be a problem. When metreon molecules are formed from metreon atoms, the number of particles decreases. Le Chatelier's principle tells us that such a reaction is favored at high pressures. At low pressures, the equilibrium favors an increase in the number of particles. For this, metreon molecules would have to be broken down into metreon atoms. However, no energy is released when chemical bonds are broken. Rather, heat must be expended for this. It would therefore not get hot, but cold. An explosion cannot be brought about in this way. The theory of a nuclear explosion is therefore probably much more likely than a chemical reaction of molecules.

Regardless of whether the explosion of the interstellar metreon gas is triggered by neutron radiation or by a chemical reaction, the question arises as to why this has not happened much earlier. A cosmic cloud consisting of an (in whatever form) unstable substance that can be easily ignited would hardly have survived for millions of years. The same would of course apply to an ignitable

---

[30] Nuclear fusion, i.e., the merging of atomic nuclei, seems unlikely in the case of metreon. Metreon isotopes are apparently very heavy atomic nuclei. This can be said because all light elements are already known. Atomic numbers 1 to 92 are already occupied by naturally occurring elements on Earth. After that, there are over twenty more artificially discovered elements. The largest natural elements are already very heavy. The artificial ones are significantly heavier. Each additional element must be much heavier. While light elements tend to fuse (even if it is still difficult to achieve), heavy elements tend to fission. Therefore, only fission seems plausible for such a heavy element as metreon.

cloud of hydrogen and oxygen. The Riker Maneuver would fundamentally be conceivable, however. Because first, metreon gas is collected and then released again in increased concentration. This relatively dense metreon gas would quickly disperse in space due to its (compared to the surroundings) high pressure. However, since the Son'a fire their weapons within a few seconds, it may be quite realistic that they ignite the cloud because it still has a fairly high density. ◄

2.4 One Moon Circles

Another example of the use of hydrogen as an explosive can be found in the *series Star Trek: The Next Generation.* In the 17th episode of the 4th TNG season, *"Night Terrors,"* almost the entire crew of the Enterprise begins to hallucinate. Only two crew members are exempt from this. One is, of course, Data. As an android, his positronic brain is naturally less susceptible to hallucinations and other delusions. The second non-hallucinating crew member is the half-Betazoid Deanna Troi. However, she has strange dreams in which she repeatedly hears a voice talking about "one moon circling."

What was going on? The Enterprise was searching for the USS Brattain. When they finally found the missing ship, it was discovered that the hallucinations had started earlier among its crew. Eventually, everyone went mad and killed each other. In the attempt to leave the corresponding region of space, the Enterprise faces the same fate as the Brattain. Initially, it is stuck. As it turns out after some time, the Enterprise and the Brattain are trapped in a so-called Tyken's Rift. A Tyken's Rift is a space anomaly named after the Melthusian captain who was the first to be trapped in one. The only way to free oneself from it is by creating a massive explosion. Massive here really means massive, as even the Enterprise's photon torpedoes are not sufficient.

However, the Enterprise (and the Brattain) are apparently not the only ships trapped in the said Tyken's Rift. Unnoticed by the Starfleet ships, another ship seems to be stuck at the other end of the rift. We do not learn much about these strangers. We only know two things: They are telepaths and need hydrogen to trigger the freeing explosion. Their telepathic signals are apparently undecipherable for most humanoid species and only disrupt their REM sleep. Therefore, the crew begins to hallucinate and go mad sooner or later. Only the half-Betazoid Deanna Troi receives the message correctly. However, she can initially make little sense of it.

Instead of a clear statement of what they actually need, the strangers repeatedly tell Deanna in her dreams that "one moon circles." After much deliberation, the correct conclusion is finally reached. It is supposed to be a reference to hydrogen. One imagines a hydrogen atom as a planet around which a moon circles. Everything is, of course, greatly reduced in size. The image is based on the Bohr model of the atom: a nucleus around which a single electron "moon" circles.

2.4 One Moon Circles

Based on this insight, the Enterprise then releases hydrogen into the rift. This is then used by the strangers to create the freeing explosion.

Why the telepathic strangers cannot express themselves more clearly and instead have to resort to the metaphor of the circling moon is another matter. For dramatic reasons, it probably makes some sense. However, the question is whether the metaphor is even an accurate description of what they need?

According to the Bohr model of the atom, a hydrogen atom is indeed a nucleus around which a single electron circles. Whether the Bohr model of the atom is a truly accurate description of reality is another matter. We will address this question again in another chapter. For hydrogen atoms, it might even be a quite useful approximation. However, the image of the circling moon still does not really fit.

First of all, what is a moon? A moon is a celestial body that circles a significantly larger celestial body. Niels Bohr would agree with this analogy for the hydrogen atom to this extent. However, a moon does not circle just any celestial body. It circles a planet. A planet, in turn, is a celestial body that circles another, significantly larger celestial body. A moon is thus a celestial body that circles a planet that circles a sun. A somewhat strange image for a hydrogen atom. In a hydrogen atom, after all, only one electron circles a nucleus. If one analogizes the term moon for the electron and the nucleus as a planet, then the question arises, where is the sun? A more fitting metaphor would therefore probably be "one planet circles." If one makes communication so difficult to understand that vital information is conveyed in the form of dreams and no clear statements are made, but metaphors are used, then at least the metaphors should be appropriate.

"One planet circles" comes closer to the reality of a hydrogen atom. But do the strangers actually want hydrogen atoms? Hydrogen does not occur in the form of hydrogen atoms but as hydrogen molecules. Two hydrogen atoms form a hydrogen molecule. Elementary hydrogen is therefore referred to as H_2 in chemistry. It is indeed possible to produce individual hydrogen atoms. However, creating and storing a gas consisting of these atoms is not feasible. Individual hydrogen atoms are called hydrogen radicals for a reason. The term radical comes from their high reactivity. They have a very high tendency to engage in chemical reactions with other substances. When a radical encounters a molecule or even another radical, it tends to immediately form a bond. In a gas of hydrogen atoms, radicals would constantly encounter each other. The gas would therefore not remain a collection of individual atoms, as the encountering radicals would immediately form a molecule. It would transform into a collection of hydrogen molecules in no time. Thus, the image of a single moon/planet circling is again inappropriate.

A hydrogen molecule consists of two hydrogen atoms. Accordingly, not one but two electrons belong to the molecule. In fact, one is dealing with two orbiting planets. However, the aliens from the other end of the Tyken's Rift did well not to simply communicate "two moons circle." That would have had to be interpreted as an image for helium. If the Enterprise had introduced helium into the rift, the alien spaceship on the other side of the Tyken's Rift would have been able to do absolutely nothing with it. There is nothing in the entire universe that is less suitable for inducing a reaction than helium.[31] A hydrogen molecule, on the other hand, consists of two "planets" orbiting two "suns." After all, there are two atomic nuclei in the molecule. The message "two planets orbiting two suns" would have been much more precise. Whether it would have led to a correct interpretation as a hint towards hydrogen more quickly is another matter.

> **A few more details**
>
> Strictly speaking, the two planets called electrons do not really orbit freely around the two suns or each of the two suns. In the hydrogen molecule, the two atomic nuclei share the two electrons in a certain way. This is the cause of the bond between the atoms. If the electrons were simply orbiting the two centrally arranged atomic nuclei or alternatively each orbiting their respective atomic nucleus individually, there would be no bond.
>
> To achieve a bond, the electrons must be located between atomic nuclei. The two positively charged atomic nuclei (which would repel each other) attract the negatively charged electrons. This attraction between atomic nuclei and electrons ultimately holds the atomic nuclei together. However, the electrons must not move freely in an orbit but must be located between the atomic nuclei. Or more precisely: They must preferably be located between the atomic nuclei. Partly they are in an orbit, partly they concentrate between the atomic nuclei. This is the basis of the so-called covalent bond and cannot be really well explained with the Bohr model of the atom.
>
> Indeed, it would be somewhat difficult to communicate this telepathically in a catchy way that fits into a dream. That is why the aliens from Star Trek probably used the simplified image of the single orbiting moon. ◄

In any case, the Enterprise draws the right conclusions. So, they send hydrogen through the Bussard collectors into the interior of the Tyken's Rift. A hydrogen cloud moves in the form of a red glowing beam from the ship towards the anomaly. Is that actually possible just like that?

---

[31] Chemically speaking, helium is a noble gas and as such is not capable of chemical reactions. This is also true for the other noble gases. For some of them (krypton and especially xenon), it has already been possible to carry out reactions in the laboratory and thereby produce noble gas compounds. This works all the better the further down the noble gas is in the periodic table of elements. For radon, the lowest naturally occurring noble gas, there is still relatively little knowledge due to its radioactivity. For the xenon above it, the production of chemical compounds works at least under laboratory conditions with great effort. Krypton, which is above that, can still be converted into an unstable compound with fluorine with the greatest effort. So far, it has only been possible to produce something like a compound from the argon that follows further up at less than minus 260 °C. With neon, it becomes even more difficult, and with the highest noble gas, helium, one does not even need to think about a reaction under laboratory conditions, let alone an explosive reaction.

First of all, the question arises as to why the hydrogen glows red. Actually, hydrogen is a colorless gas, isn't it? In principle, that is correct. With the naked eye, one cannot distinguish hydrogen from air. Both are colorless gases. But that does not mean that hydrogen cannot have any colors. When atoms are energetically excited, it can happen that electrons temporarily move to a higher energy level. When they return to a lower energy level, they emit light. The wavelength of this light is inversely proportional to the difference between the two energy levels and specific to the respective element. Depending on how far apart the individual energy levels are in an element, the emitted photons have different amounts of energy. The energy of the photons corresponds to a certain wavelength of light and thus a color.

Hydrogen, like all elements, has several possible energetic states and correspondingly several wavelengths that correspond to the respective differences in energy levels. When a hydrogen atom returns to its lowest energy state (the K-shell), the most energy is released. The resulting light is therefore very energetic and correspondingly short-wavelength. It is not visible to the naked eye because it belongs to the ultraviolet part of the spectrum. If the hydrogen atom first returns only to its second-lowest energy state (the L-shell), less energy is also released. The resulting light is correspondingly longer-wavelength and now actually in the visible range of the spectrum. Depending on which excited state the atom returns from, the amount of energy released still differs a bit. If it returns from a very high energy level, the released light is blue or even violet. However, if it has only moved from the third-lowest energy level (the M-shell) to the second-lowest energy level, light with a wavelength of 656.28 nanometers is emitted. This characteristic wavelength belongs to red light and is known as H-alpha.[32] H-alpha is indeed the brightest visible spectral line of hydrogen. A red glow would therefore not be entirely far-fetched for excited hydrogen.

However, the question remains as to what makes the hydrogen glow. Hydrogen does not start emitting light on its own. It must be somehow energetically excited to glow. Perhaps answering the next question will help us further.

Another problem with introducing hydrogen into the interior of the Tyken's Rift is the direction of movement. If hydrogen is released into space, it distributes evenly in the vacuum. It does not move as a beam directly to where it is supposed to go. If the hydrogen exits at a very high speed, its molecules have a certain

---

[32] In addition, hydrogen has a whole range of other characteristic wavelengths for the light it emits. These are called spectral lines. Spectral lines are characteristic because they occur with their specific wavelengths only in hydrogen. Every other element also has spectral lines, but these are again specific to the respective element. This effect is used not only in chemical analysis but also in astrochemistry. It involves determining the chemical composition of distant celestial bodies (which is exciting in itself because it could reveal whether extraterrestrial life is fundamentally possible in another part of the universe). Since one cannot simply fly to distant exoplanets to take a sample (at least not until Zefram Cochrane will finally invent the warp drive), such investigations can only be conducted based on light that has been emitted or absorbed by the corresponding substances.

preferential momentum. The momentum corresponds to the product of the molecules' mass and their speed. The principle of conservation of momentum states not only that the numerical value of the momentum is conserved but also that the direction of movement is conserved. In an elastic collision of two bodies, the direction of movement of the individual bodies can change, but the total momentum (considering the direction) remains conserved. So if the hydrogen atoms exit the spaceship at high speed in the direction of the Tyken's Rift, they retain this direction of movement. In the vacuum of space, there is nothing to stop them. However, the exit direction from the Bussard collector is only a preferential direction. It is by no means the only direction of movement that the molecules have.

In a gas, all molecules move randomly in all directions. In a fast gas stream, this movement of the molecules inside is superimposed by the external movement of the gas stream. There is a kind of preferential direction. Nevertheless, hardly any molecule moves exactly in the direction of the target but a bit away from the axis of the imagined hydrogen beam. This is not a problem as long as the gas flows through a pipe. Once a gas molecule reaches the "edge" of the gas beam, it hits the pipe wall. The collision with the pipe wall redirects it back into the interior of the gas beam. Once the gas beam has left the spaceship and thus the pipe, this no longer applies. If a molecule reaches the edge of the gas beam, there is nothing to stop it. It would simply continue to move away from the axis of the beam. Therefore, one would not observe a beam of hydrogen exiting the spaceship but a cone. The faster the gas exits, the more pointed the cone becomes. However, a widening of the beam into a cone would be unavoidable. In the case of the hydrogen beam emitted by the Enterprise into the Tyken's Rift, we do not see this. Instead, we see a well-defined beam with a constant thickness.

The Enterprise must therefore do something to direct the hydrogen and prevent its spread into the vacuum. This could ultimately also explain the glow of the hydrogen. Since the hydrogen can no longer be enclosed by the wall of a pipe outside the spaceship, this seems to be done by some kind of force field. The hydrogen is expelled through the Bussard collectors. These are actually designed to collect gases like hydrogen from interstellar space. This requires suitable force fields. We do not know exactly how this works (after all, the Bussard collector has not yet been invented). However, energy must somehow be transferred to the hydrogen beam to prevent its widening into a cone. If energy is transferred to the hydrogen, the hydrogen atoms are energetically excited. The result could be the observed red glow of the hydrogen, as part of the excitation energy is released in the form of light.

If it is possible in this way to direct hydrogen into the interior of the Tyken's Rift, the aliens can cause an explosion with it. But is this explosion sufficient to dissolve the rift? Since we do not know much about (the yet undiscovered) Tyken's Rifts, we cannot say at first how much explosive power is needed to destroy them. However, igniting hydrogen seems hardly suitable to provide enough energy. Why can we say this when we do not know how much explosive power is needed?

2.4 One Moon Circles

Well, what we learn from Commander Data is that a photon torpedo does not have enough explosive power. Even if we do not know the exact amount of energy released during the detonation of a photon torpedo, we at least know the basic principle of its function. Photon torpedoes are a weapon technology from Star Trek, where large amounts of energy are released suddenly by the reaction of antimatter with matter. When antimatter meets matter, both are completely converted into energy. This means that all mass is converted into energy. Nuclear weapons are also based on the conversion of mass into energy. However, only a small part of the mass is converted into energy. When antimatter and matter collide, the entire mass of both is completely converted into energy. Therefore, an unimaginable amount of energy is released during the detonation of a photon torpedo. If this amount of energy is not enough to destroy a Tyken's Rift, then it really takes a lot of energy. The question arises as to how a oxyhydrogen explosion is supposed to provide this amount of energy.

If the Enterprise were to direct 1000 tons of hydrogen into the Tyken's Rift (which would be an enormous amount) and the aliens on the other side were to mix it with 8000 tons of oxygen[33], then 120 gigajoules of energy would be released in the explosion. That would already be an enormous amount of energy. One should not get too close to such an explosion. But how much is that compared to a photon torpedo?

The answer is: almost nothing. With Einstein's famous equation $E = m \cdot c^2$, we can convert the amount of energy released into a "disappeared" mass. This mass then corresponds to the total amount of antimatter plus matter used in the photon torpedo. Since the speed of light c is very large, it does not take much mass m for this amount of energy. We arrive at not even 1.4 milligrams. This means that a photon torpedo with just 0.7 milligrams of antimatter could already produce such an explosion. Even without knowing their detailed specifications, one can probably assume that Starfleet's photon torpedoes in the 24th century have more explosive power. Besides, it seems somewhat unlikely that the spaceship on the other side has 8000 tons of oxygen on board and can use it to dissolve the Tyken's Rift. Moreover, it should be considered that the two gases must be well mixed beforehand for a proper explosion. Even though I am very reluctant to contradict Data's calculations: the photon torpedoes would probably have been much more promising than the hydrogen explosion. Only then the episode would have been over after just a few minutes.

---

[33] For a stoichiometric combustion, one oxygen molecule must come for every two hydrogen molecules. However, one oxygen molecule weighs 16 times as much as one hydrogen molecule. Therefore, although only half the amount of oxygen molecules is needed, the mass of oxygen required is eight times greater.

Atoms in a Completely Different Way 3

3.1 When Atoms Burn

After three years aboard the *Voyager*, the Ocampa Kes, who had until then supported the medical-holographic emergency hologram as a nurse, left the spaceship. This departure was quite impressive because it was accompanied by the full development of her telepathic powers. She is supported on her path to developing her abilities by the Vulcan Tuvok. Vulcans are known to possess telepathic abilities themselves, although these are not nearly as impressive as those of the Ocampa. Vulcans must not only touch the other person for their famous mind meld but are also primarily limited to telepathy. As a teacher on the complicated (and not entirely safe) path of telepathy, a Vulcan with his disciplined mind still seems quite suitable. In the case of Kes, however, it turns out that Ocampa are capable of much more than even Tuvok can imagine. This includes, among other things, telekinetic abilities. Ocampa are not only able to read the thoughts of other people and communicate telepathically. They are even able to move things with the power of their thoughts.

In the 2nd episode of the 4th VOY season, *"The Gift"*, Kes's abilities finally come to full fruition. She not only displays some physically remarkable abilities but also demonstrates that she is far beyond our understanding of chemistry.

During one of her sessions with Tuvok, Kes begins not just to see the candle in front of her with her eyes. Her gaze expands to the point where she sees the fire itself with her mind. In doing so, she sees what a chemist thinks of when they hear the word combustion: She begins to perceive the individual molecules moving quickly and reacting with each other. Then she starts to look even deeper and actively intervene in the chemical process itself—on a subatomic level. Finally, her mental gaze even extends beyond the subatomic, which Tuvok then considers somewhat far-fetched.

Let's stick to what she does with the flame. She intensifies it, making it significantly larger. As mentioned, Kes makes this intervention in chemistry on a subatomic level. How is this to be imagined, and does it make any sense chemically?

First, let's look at combustion itself chemically. Combustion can be simplified as a rapid total oxidation. Rapid means that we are not talking about a slow, creeping process like the rusting of a nail, even if it may be a chemically quite similar reaction. Total oxidation means that the (in this case organic) fuel is converted to carbon dioxide and water. In other oxidation reactions, the reaction only runs partially. Then, for example, alcohols or carbonyl compounds would be formed.

In practice, total oxidation does not necessarily mean that the entire fuel is actually converted. Especially in the case of oxygen deficiency, part of the fuel molecules is not or only partially burned. Incompletely burned fuel molecules usually aggregate into small particles known as soot. Since they are often very small, they can penetrate deep into the lungs when inhaled and cause damage there. These soot particles are one of the main causes of the problem often discussed under the term fine dust. In clean combustion, as little unburned fuel as possible should be carried away from the flame in the form of soot particles and distributed in the room.

When the fuel is converted with oxygen, it is still not necessarily a total oxidation. In an ideal combustion, the fuel molecules are completely broken down and converted with oxygen. The hydrogen atoms are converted with oxygen to water. The carbon atoms are converted with oxygen to carbon dioxide. And the oxygen atoms that were already part of the fuel molecule also become part of water or carbon dioxide molecules. The more oxygen already contained in the fuel molecules, the less atmospheric oxygen is logically needed for combustion.[1]

In reality, the oxygen available in the flame is not always sufficient to completely convert all the carbon to carbon dioxide. Soot particles, which essentially consist of unburned carbon, are not the only possible byproduct of incomplete combustion. Another possibility is the formation of carbon monoxide CO. In this case, a carbon atom is not converted with two oxygen atoms as in carbon dioxide CO_2, but only with one. This is not only unfavorable because a significant portion of the energy is released only in the final reaction step (the complete oxidation of CO to CO_2). Carbon monoxide is particularly problematic because it is very toxic.

---

[1] An important example of a fuel that contains oxygen is ethanol, which is added to gasoline as a biofuel. Combustion-wise, oxygen-containing fuel molecules have both advantages and disadvantages. One advantage is that they often burn cleaner. Less soot is formed, and the fine dust problem is therefore less pronounced. A disadvantage, however, is that the molecules are somewhat pre-oxidized. The first step of combustion has already taken place, so to speak. Consequently, the energy released in this first reaction step of oxidation can no longer be utilized. The energy content of oxygen-containing fuels is therefore lower. Another disadvantage is that the molecules are more polar than pure hydrocarbons. As a result, more water can dissolve in oxygen-containing fuels. This water increases the weight and volume of the fuel but not its calorific value. Moreover, it must be vaporized during combustion. This requires energy, which reduces the effectively usable energy of combustion.

3.1 When Atoms Burn

In the blood, it binds to hemoglobin like oxygen molecules—only much more strongly. As a result, the hemoglobin is occupied, and less oxygen can be transported from the lungs to the body.

When looking at a flame from the outside, the chemistry within it can usually be sufficiently described for most questions by the following reaction:

$$\text{hydrocarbon} + \text{oxygen} \rightarrow \text{carbon dioxide} + \text{water}$$

Hydrocarbons can be all possible alkanes, alkenes, alkynes, and aromatics here. Additionally, in our case, oxygen-containing molecules such as alcohols or fatty acids are also meant. If one wants to be more precise, then one should also take this already described side reaction into account:

$$\text{hydrocarbon} + \text{oxygen} \rightarrow \text{carbon monoxide} + \text{water}$$

Otherwise, one should be aware of the fact that parts of the fuel often do not burn properly, which leads to soot formation. In many cases, this is sufficient as a chemical description of combustion. However, it becomes more complex if one wants to engage in advanced combustion technology. Or if one, like Kes, wants to influence combustion telekinetically on a subatomic level. For this, one must be aware that the reactions described above do not actually exist in that form.

Let's take a look at a candle flame. Before the chemical reactions can begin, the wax must first evaporate. In the process, the first hydrocarbon molecules begin to break into smaller fragments due to the high temperature. At the latest, upon reaching the actual flame, the molecules break apart completely. In the process, bonds are broken. The resulting fragments often have single electrons attached to them. These can be thought of as half bonds. These half bonds have a very strong tendency to combine with other half bonds, forming new molecules. Due to this strong reactivity, these intermediates are called radicals (which we have already encountered above). The various, differently sized molecules combine with each other or with oxygen molecules (or individual oxygen atoms) to form other intermediates. Most of these intermediates have a lifespan of only fractions of a second. The whole process is so complex that the reaction mechanism of something as simple as a candle flame has not been fully elucidated to this day. However, the processes in a candle flame are now largely understood. The result is that one arrives at a three-digit number of individual reactions. It begins with the aforementioned breaking of the large fuel molecules into smaller fragments.[2] These react with each other or with oxygen or with oxygen-containing intermediates or... Well, in the end, carbon dioxide and water are essentially produced. In most cases, it is sufficient to consider this global reaction. However, not if one wants to influence combustion on a subatomic level.

---

[2] Given the fact that wax is not just a single, pure substance but a mixture of various hydrocarbons, and each of these hydrocarbons can break down into fragments in different ways, an incredible number of individual reactions arise from this step alone.

The above description refers to the molecular level. The submolecular level would then be the atomic level (after all, molecules are made up of atoms). The subatomic level would be the level of atomic nuclei and electrons. Wanting to influence combustion on this level makes sense in a certain way. We have already mentioned the bonds between atoms and the half bonds that we call radicals. The bonds between atoms are mediated by two electrons each. In radicals, there are single electrons that seek a partner to form a bond together. Essentially, combustion (like any chemical reaction) is based on the breaking and forming of chemical bonds. The electrons are, simply put, responsible for the formation of bonds and radicals. Changing these is one way to influence combustion. If Kes can actually manage to specifically shift electrons within the molecules with her telekinetic abilities so that, for example, bonds break, then she can indeed accelerate the chemical reaction. In a way, she would be influencing combustion on a subatomic level.

On the screen, however, it looks more like she is accelerating the atoms and molecules. This would mean her intervention takes place on the atomic or molecular level.[3] This would be just as conceivable. Possibly even somewhat simpler, as one would not have to operate on such a tiny scale. Nevertheless, the scale would still be incredibly small. Could one manipulate a chemical reaction without influencing the electrons that are crucial for chemical bonds?

In order for two molecules to react with each other, they first need to meet. In the flame, the molecules move more or less randomly back and forth. This results in frequent collisions. If it were possible to influence the movement of the molecules, one could increase the number of collisions. This could intensify the combustion. However, merely having the molecules meet is not enough. It can certainly happen that molecule A hits the wrong part of molecule B. The probability of an actual reaction could be significantly increased if it were possible to always orient the molecules correctly. The strongest effect would likely be achieved by increasing the speed of the molecules. The higher the speed, the greater the energy that can cause a reaction upon collision.

To increase the speed, one would need to raise the temperature. This can be achieved in two ways: by adding heat or by preventing the flame from dissipating heat. In principle, this would be the simplest approach. However, it would ultimately have little to do with an atomic (or even subatomic) influence on the flame, as Kes does. How it could be physically realized is another matter entirely.

Nevertheless, it would be conceivable to increase the speed of the particles without having to add energy. The molecules move at different speeds. Simply put, there are relatively few slow molecules, many medium-speed molecules, and again relatively few fast molecules. The higher the temperature, the higher the average speed.

---

[3] However, it is also conceivable that the increased speed of the atoms seen in the episode is the result of subatomic intervention by Kes. If she managed to accelerate combustion through subatomic interventions, this would ultimately lead to a higher temperature. As a result, the atoms and molecules would move correspondingly faster.

At a high temperature, the proportion of sufficiently fast molecules for a reaction is greater. However, even at low temperatures, there are always a few molecules that are fast enough. The proportion is just significantly lower. That is why a fuel only ignites when it reaches a certain temperature. Even if isolated reactions occur between molecules, the heat from the reactions is dissipated too quickly to the surroundings. As a result, the system cannot heat itself up, and the reaction speed cannot increase. It is not that there are three groups of molecules: slow, medium-speed, and fast. All speeds occur, just with different frequencies. Or, in relation to the individual molecule: with different probabilities. Now, let's imagine two molecules, each with a speed that is too low for a reaction. If Kes could redistribute kinetic energy between the molecules (let's ignore the issue of momentum conservation here), one could at least imagine that almost all the energy would be transferred to one of the two molecules. This molecule would then be very fast and could trigger a reaction upon collision. The other would then be very slow. But that ultimately makes little difference. Its energy was already insufficient for a reaction before. In this way, one could theoretically intensify the combustion without having to add energy.

So much for the theory. Now we come to the real challenge. How does one telekinetically influence individual atoms, entire molecules, or even (on a subatomic level) electrons? Just like with telepathy (often encountered in Star Trek), we do not really know how the brain of an Ocampa is supposed to influence things outside their body.

It is likely to be particularly challenging to control things on a (sub)atomic level. To be able to do that, one must first capture the corresponding thing. One must, in a sense, see the atoms, molecules, or electrons before one's inner eye. The minimum requirement would be to be able to recognize the location and speed of the molecules.[4] Additionally, it would be at least very helpful if one could recognize their nature. Ultimately, it is simply about knowing what kind of molecule it is. So, one should not only recognize that something is there but also what it is. We do not know exactly what the corresponding sensory organ should look like. However, it is clear that a sensory organ can hardly capture things that are smaller than the cells it is made of. Biological cells are microscopically small.[5] Compared to atoms and molecules, however, they are huge. And that's where the problems begin. Kes does not "see" atoms in her brain, but rather distant atoms. One might argue that Tuvok's meditation lamp is directly in front of Kes and thus anything but far away. Chemically speaking, however, a meter is an unimaginably

---

[4] The Heisenberg uncertainty principle, which we will get to know later, would come as an additional challenge at this point. However, we do not want to consider that further here for now.

[5] This example again illustrates that even recognizing things that are as large as the corresponding sensory cells is not possible. At least not if one wants to recognize spatially resolved where the thing is or even what it looks like in detail. However, molecules can fundamentally be perceived. That is exactly what our two chemical senses (taste and smell) do. These senses can only fundamentally recognize that the corresponding molecules are there. Moreover, the molecules must come to the body. The sense of smell is not a remote sense that could perceive molecules without direct contact.

large distance. Not only measured by the size of the atoms but also measured by the significantly larger distance between the molecules, a meter is an incredible distance. The problem becomes clear when one realizes what lies in between. Even if the Ocampa have a corresponding sensory organ outside the skull (which at least saves having to look through the rather compact bone), there is still about a meter of air in between. The particle density in gases may be relatively low. Nevertheless, about 30 trillion molecules are found in a liter of air under standard conditions. The proverbial search for a needle in a haystack is probably no longer an apt metaphor for this. It is much worse. Capturing individual molecules through so many molecules, amidst an unimaginable number of other molecules, is... well... let's say challenging.[6]

Even if one gets the "seeing" of atoms under control, the difficulties continue. Next, one must influence them. When we humans influence any things outside our bodies, we touch them. Biology also knows a few chemical methods in which substances are released that influence things outside the body or cell. None of this happens with telekinesis. Here, things must be influenced remotely without contact, without adding another substance. In our case: molecules must be influenced in their movement from a distance. Can that even work?

In the case of oxygen molecules, it would indeed be difficult. These molecules, essential for combustion, are completely electrically neutral. Not only do they have no net charge, but within the molecule, electrons and atomic nuclei are distributed in such a way that there are no local charge concentrations. In chemistry, such a molecule is called nonpolar. It would be difficult to influence this molecule without contact. However, this does not apply to all molecules that occur during combustion. Many of the intermediate products are indeed polar. This means they have a positive charge at one end of the molecule and a negative charge at the other end. In total, they are electrically neutral. However, charges are present on a small scale. One could at least theoretically do something with that. Additionally, individual ionic species occur during combustion, meaning individual, albeit short-lived, molecules are actually electrically charged. These so-called ions can be influenced even better. On a subatomic level, we finally find the electrons. They are negatively charged and could thus—at least in principle—be influenced and shifted within the molecule. This would have an enormous impact on the chemical bonds.

To influence charged particles from a distance, there are two related but different effects that can be used. One could either apply an electric or a magnetic field. With an electric field, one can move a charged particle directly in one direction. With a magnetic field, one can at least influence moving charged particles.

---

[6] Some might argue at this point that one can see through a meter of air without any problems. For a sensory organ that captures relatively large objects optically, like the human eye, this is indeed not a problem. For a sensory organ that "sees" individual molecules (in whatever way), the molecules of the air are probably a problem.

In principle, it is not particularly difficult to generate an electric or magnetic field. For an electric field, for example, you only need to charge one metal plate positively and another metal plate negatively. An electric field then forms between them. A magnetic field can be generated with a coil. To do this, you wind a conductor into a spiral and pass an electric current through it. You already have a magnetic field. Technically, this is all no problem. Biologically, however, it is considerably more difficult. Although living beings can generate electric current—our nerves are based precisely on this—not much current would be needed to influence individual (charged or at least polar) atoms. However, Kes influences quite a lot of atoms simultaneously when she seriously enhances the combustion in the flame. Correspondingly, quite intense currents would have to flow through an Ocampa brain, and a variety of different currents would be needed to influence various molecules simultaneously. A human brain would hardly survive this (and could simply not generate them). But Kes demonstrates quite clearly that Ocampa brains can achieve significantly more than human brains.

The real challenge, however, lies elsewhere. This problem is ultimately closely related to the problem of "seeing" the molecules described above. To influence combustion with the power of thought, one must control very many, very small, and very fast particles simultaneously. How can one specifically influence only the atoms in the flame when countless trillions of atoms lie between one's own brain and the flame? A brain (which differs significantly from a human brain) might be able to generate an electric or magnetic field. Extending this field significantly beyond the area of the skull then becomes considerably more difficult. Focusing the field on a specific area (the flame) far away from the brain makes the matter even more challenging.

And the main problem is ultimately something entirely different. There are indeed chemical reactions driven by positively charged ions being pulled to one side and negatively charged ions to the other side in an electric field. This is called electrolysis. Influencing combustion in the way Kes does, however, is something entirely different. To intensify combustion, one must specifically bring together molecules that otherwise move randomly with their reaction partners. One must influence their direction and speed. One thing must be considered: the correct direction is different for each particle. Unlike electrolysis, not all particles should move in the same direction. Each particle must be directed in a different direction. Therefore, the electric or magnetic field would have to be oriented differently for each molecule. Every few nanometers, it would need a different orientation. And that's not all. As soon as one particle is directed to its target, a new particle enters the same spatial area. Who says this particle should be directed in the same direction? The orientation of the electric or magnetic field would therefore need to be changed—within fractions of microseconds or rather nanoseconds. To do this, all molecules would have to be observed individually, encountering the problems described above. Based on this observation, the telekinetic would have to make billions of decisions about the reorientation of local electric fields within absurdly short times and implement these decisions. Without wanting to offend Kes: This seems (to put it mildly) very demanding.

Considering combustion on a molecular, atomic, or even subatomic level at the individual electrons would certainly be very exciting. However, wanting to influence it directly on this level is far beyond anything that makes our understanding of molecular processes seem realistic. Ultimately, the approach fundamentally differs from what chemistry is in practice. In chemistry, one must understand what happens with individual molecules. However, one almost never handles individual molecules but always many trillions or quadrillions simultaneously. In a chemical laboratory, so many molecules are always added simultaneously that they are not counted but weighed. Accordingly, no one in a chemistry lab would think of influencing the molecules individually. Instead, one changes the temperature or pressure, adds a solvent, supplements a catalyst to accelerate the reaction, or introduces a reactant to the reaction. This is ultimately what a practically-minded chemist would do to intensify the flame. Instead of trying to influence the individual atoms with the power of thought, one would rather try to improve the air supply to the combustion.

3.2 Tiny Atoms

People who are shrunk to the size of ants or smaller are a popular theme in a large number of films. The scientific backgrounds are all too often treated very imprecisely. Star Trek makes exemplary efforts by considering aspects that are otherwise often forgotten. But what scientific questions would actually need to be considered in a shrinking ray or similar technology?

An example of this can be seen in the 14th episode of the 6th DS9 season, *"One Little Ship"*. The crew of the space station is on a research mission with the USS Defiant. To investigate a subspace anomaly more closely, Jadzia Dax, Miles O'Brien, and Dr. Julian Bashir boarded the shuttle Rubicon and flew into the anomaly. This subspace anomaly causes strong spatial distortions, which lead to objects that come too close to it shrinking significantly. To somewhat protect the shuttle and its crew from the effects of this phenomenon, the Defiant holds the Rubicon with the tractor beam. Unfortunately, we are currently in the sixth season of the series and thus in the year 2374. In other words: at the height of the conflict with the Dominion. What must happen, happens: The Jem'Hadar attack at exactly the wrong moment. It would be bad enough if they just captured and took over the Defiant. Additionally, the tractor beam breaks off, and the poor Rubicon along with its three-member crew remains in the subspace anomaly. Although the three manage to escape from it, there is a problem. They have not grown back to their original size. In hindsight, this annoying mishap turns out to be a lucky coincidence, as it helps to drive the Jem'Hadar off the Defiant again. And of course, by the end of the episode, the shrinking process is reversed. But initially, it presents the three on the Rubicon with enormous challenges and also raises some scientific and chemical questions for us.

First of all, the question arises as to what actually happens to the atoms and molecules. Miniaturization in today's technology means producing something in

3.2 Tiny Atoms

a smaller format. For example, if you produce a model of a real object at a scale of one to ten, you use one-thousandth of the material that you would use for the original.[7] But can the amount simply decrease when shrinking in a subspace anomaly? Or put differently: What would happen biochemically then? Since the Rubicon and its crew look the same as before (just significantly smaller), nothing could have simply been taken away from one end. Rather, the removed material must have been evenly removed from the inside of the body. If it did not simply dissolve into nothing[8], then it must have somehow escaped from the body. Since Dax, O'Brien, and Bashir did not explode when shrinking, which would inevitably be the consequence, this obviously did not happen.

But even if the excess molecules simply disappear, problems arise. Because all these molecules have a function in the organism. Let's imagine a biological cell. There is the DNA in the cell nucleus[9], where the genetic information is stored. In addition, there is an incredibly large number of proteins that have a precisely defined molecular structure, which enables them to fulfill their function. If you were to remove atoms evenly from the body (and when shrinking a 1.70 m tall person to the size of 1 cm, we are talking about removing 99.99998% of all molecules), then only a few atoms would remain even from a very large protein. So, almost nothing would be left of the protein molecule. This would destroy the structure of the molecule, its function would be lost, and Dax, O'Brien, and Bashir would die instantly. Even if the shrunken organism somehow managed to survive the failure of, for example, the enzymes for a short time, they could not be newly formed. First of all, the disappearance of most atoms would certainly have destroyed the ribosomes, where proteins are formed. Moreover, the genetic material, where the blueprints of the proteins are stored, would be just as little left because the DNA would suffer the same fate as the proteins.

The creators of Star Trek were apparently aware of this problem. So, the atoms on board the Rubicon do not simply disappear, but they become smaller. But what would be the consequences of that?

In science, there are a few immutable principles. One of them is the conservation of mass. No matter what happens: The mass remains the same. A person can indeed reduce their mass (otherwise diets would be pointless). Nevertheless, the total mass remains constant. Although the mass of the person changes, the mass of the universe remains the same. Atoms that were previously in our fat cells are

---

[7] If you scale down an object so that the original object with a diameter of, for example, 1 m measures only 10 cm, the mass is reduced to one-thousandth because the volume (and with constant density, also the mass) scales cubically with the diameter. That means if you multiply the length by a factor of one-tenth, then the volume is multiplied by a factor of one-tenth cubed; in other words: one-thousandth.

[8] The German expression "in Luft aufgelöst" (dissolved into thin air) would not fit here either, as air is always matter and, moreover, the volume does not decrease when transitioning to the gas phase but increases significantly.

[9] The abbreviation DNA stands for *deoxyribonucleic acid*.

simply somewhere else afterward. The atoms have not become smaller and lighter. The process of weight change in humans also works in the other direction, as most of us know from painful experience. However, the same principle applies again. The additional mass does not come from nothing but was previously in the things we ate. The change in mass (and thus the number of atoms) is always equal to the mass (or number of atoms) added—for example, as food—minus the mass/number of atoms removed:

Change in the number of atoms in the cell = number of atoms added—number of atoms removed.

If the atoms were to shrink, it would mean that they have transformed into some (as of yet unknown) new particles. For example, the transformation of neutrons, which are found in atomic nuclei, into other elementary particles is a well-known process. In the so-called beta decay, it transforms into three particles at once. A proton is formed, and in addition, an electron (the beta particle) and a neutrino (or more precisely: an antineutrino). If one were to measure the mass of the new particles, one would find that it almost corresponds to the mass of the decayed neutron. The only tiny difference is that the rest masses of the proton, electron, and neutrino together are slightly smaller than the mass of the neutron. However, the electron and neutrino are very fast and thus have considerable kinetic energy. This energy closes the mass balance again. Therefore, the total mass remains constant. The difference in masses and their compensation through energy is described by a very well-known equation: $E = m \cdot c^2$ from Einstein's general theory of relativity.

Ultimately, the equation says nothing other than that energy E and mass m are equivalent. If energy is released in a process, the mass seemingly decreases. However, since the released energy is equivalent to the mass, the mass somehow remains conserved. Since the speed of light (denoted by the parameter c) is very large, very little mass corresponds to a lot of energy. Therefore, the change in mass is usually not noticeable at all. When heat is released in a chemical reaction such as combustion, the corresponding amount of energy can be converted into mass, and it is found that it is almost nothing. When weighing the reactants and products, the result is always that the masses are equal. The difference is smaller than the measurement accuracy. An example where the mass differences between reactants and products are somewhat larger is nuclear processes. For this reason, nuclear fuel rods must be stored in an interim storage facility for years after use to cool down. Due to the radioactive decay processes within them, energy is continuously released because the decay products are still lighter than the heavy starting nuclides, even when the mass of the alpha particles released during decay (a chemist would say: helium nuclei) is included. Consequently, they constantly heat themselves and do not simply cool down within a few hours or days.

Similar nuclear processes, where the conversion of mass into energy occurs much faster, are known in connection with an invention from the 1940s: the atomic bomb. The fission products have a significantly lower mass than the reactants. Although it initially involves only a very small mass difference, multiplied

3.2 Tiny Atoms

by the speed of light squared, it results in a large amount of energy. And this energy is released abruptly. Just like the shrinking of the Rubicon.

In the atomic bomb dropped over Hiroshima, an amount of energy equivalent to about one gram of mass was released. If a person were shrunk to a size of one centimeter, they would subsequently weigh less than 1 g. Attention: That would be the remaining mass. Their entire remaining original mass would be released as energy. Per person, approximately 75 kg of mass would be converted into energy, depending on the specific individual. That would correspond to 75,000 Hiroshima bombs. In the case of the shrinking of the Rubicon, the mass of three adult humans[10] plus a small spaceship would have to be converted into energy. The resulting explosion would likely completely destroy even a spaceship like the Defiant from many kilometers away. Albert Einstein should better not look too closely at this part of the episode.

However, the creators of Star Trek considered an aspect that filmmakers usually completely overlook when shrinking people. After the Rubicon returns to the Defiant and flies into the ship through a plasma opening, its three shrunken crew members find that it has been boarded by the Jem'Hadar. Before they can deal with growing back to normal size, their main task is to help liberate the ship. To regain control of the ship, Chief O'Brien must manually reroute the encryption sub-processors. To do this, Dax must beam him into the corresponding circuit housing. The Chief is not thrilled with this idea for two reasons. First, he suspects that a one-centimeter man would be fried by the smallest mistake while wandering through the circuit housing. Second, Dr. Bashir points out an important circumstance: The oxygen molecules outside the Rubicon are all normal-sized. The hemoglobin molecules in the Chief's blood, however, are tiny (even tinier than "normal" molecules already are). The oxygen could not bind to his hemoglobin, so the oxygen could not be transported in the body. As a result, he would suffocate in a very short time (apart from the fact that the problems with binding the much too large oxygen molecule to the transport substance hemoglobin would continue in the biochemical processes inside the body). Fortunately, we are dealing with smart Starfleet officers. Calculating that the housing is airtight[11], Jadzia Dax first beams some air from the shuttle into the interior of the housing. This way, the Chief can breathe there for about 20 minutes and save the ship from the Jem'Hadar.

---

[10] Apologies. It is, of course, two adult humans and one Trill. For the calculation, it does not matter whether they are humans or aliens.

[11] In this case, the tightness is a bit tricky. Even if the housing is airtight, it is airtight for normal oxygen molecules. However, the molecules beamed in by Dax are many times smaller. Gases of smaller molecules tend to diffuse through even the smallest cracks or even through the material of the wall itself much more than large molecules do (Anyone who has ever tried to seal something so that hydrogen, the smallest of all real molecules, cannot escape knows the problem; with the even smaller molecules from the shrunken Rubicon, this problem will certainly be even more pronounced).

At this point, one can really be proud of Star Trek, because I don't know of any other film with a shrink ray or something similar that would have thought of this problem.

Excursus

What problems do 1 cm tall people face?
What difficulties or even advantages would people who are only 1 cm tall actually face? Let's assume that our miniature people are not shrunk like the crew of the Rubicon, but are made up of completely normal atoms and molecules. In nature, we can observe countless animals of this size class or smaller. Basically, there is nothing to prevent higher beings from being that small. But what consequences would such a small size have for a human?

One problem might be the brain. Although such a small creature can also easily have a brain, this brain would naturally be significantly smaller. Much smaller, in fact. The entire one-centimeter miniature human would, in terms of mass, not even have a ten-thousandth of the size of an average human brain. Consequently, our miniature people would probably hardly be capable of any special intellectual achievements—let alone being as smart as Jadzia Dax, Miles O'Brien, and Dr. Julian Bashir. To achieve the intellectual abilities of a human, the brains of the miniature humanoids would therefore have to be built much more efficiently. Biologically, a lot is conceivable. The human brain encompasses large areas that are little used. In fact, human cognitive abilities are performed in a remarkably small part of the brain. With a different brain structure, a lot could therefore be rationalized. Nevertheless, the brain of the miniature people would still be far too small to achieve even average intelligence.

So, could the brain be made more efficient on a smaller scale? That also becomes difficult. Biological brains are based on nerve cells and their connections to each other. These cells simply have a certain size. There is a certain range of variation, but the order of magnitude is fixed. If one were to try to cultivate brain cells that were much smaller, these cells would probably no longer be viable.

In addition, the storage of information (better known as memory) could become a problem. The brain stores information biochemically in the long term. Just as the corresponding molecules are housed in the human brain, a one-centimeter man (or a one-centimeter Trill woman) would probably have only a very limited memory capacity. However, studies on so-called savant syndrome suggest that humans are fundamentally capable of storing much larger amounts of information than most of us do. The brain structure of the miniature people would therefore have to be significantly different from ours. In principle, however, it would be conceivable that they could have the same memory performance as we do.

Another problem that miniature people would face is that the surface area of their bodies would be very large. At first glance, this seems paradoxical, and

3.2 Tiny Atoms

indeed the surface area of a one-centimeter person is naturally much smaller than that of a normal-sized person. However, relative to body volume, it is huge. As we have already seen above, the volume decreases with an exponent of three compared to the diameter when shrinking. Therefore, the volume (and with it the mass) decreases much faster than the linear dimensions. The surface area also decreases faster than the diameter, but here the exponent is only two. This means that a person half as tall (half the height) with the same body proportions would have only one-eighth (one-half cubed) of the volume. The surface area would decrease to one-quarter (one-half squared). Consequently, the body surface area relative to the volume would have increased by a factor of two.

Surface: $\frac{1}{2^2} = \frac{1}{4}$

Volume: $\frac{1}{2^3} = \frac{1}{8}$

Surface per Volume: $\frac{1/4}{1/8} = \frac{8}{4} = 2$

This game can be continued. If you imagine a 1.70 m tall person being shrunk to 1 cm, then their body surface area would have shrunk to one 28,900th (1 divided by 170 squared). However, their volume would be just over one five-millionth (to be precise, one 4,913,000th = 1 divided by 170 cubed) of the original value. Consequently, their surface-to-volume ratio would have increased by a factor of 170.

Surface: $\frac{1}{170^2} = \frac{1}{28,900}$

Volume: $\frac{1}{170^3} = \frac{1}{4,913,000}$

Surface per Volumen: $\frac{1/28,900}{1/4,913,000} = \frac{4,913,000}{28900} = 170$

Why is this important now?
In technical chemistry, the surface-to-volume ratio plays a significant role. This is because all exchange processes occur over the surface. In chemical processes, it is crucial that substances, for example, are absorbed or released by a particle as quickly as possible. The smaller the particles, the faster this happens because the surface is larger in comparison to the volume. To illustrate this with an example from daily life: This is the reason why powdered sugar exists. The same amount of powdered sugar has a much larger surface area than regular granulated sugar. As a result, it dissolves faster upon contact with water. Consequently, a larger portion of the powdered sugar dissolves in the mouth and contributes to the sweetness. "Regular" sugar, on the other hand,

dissolves more slowly because the large sugar crystals have a smaller surface area relative to their volume. Therefore, only a part of the sugar is available in the mouth to trigger a sweet taste sensation. A similar effect would affect our miniature humans. Humans evaporate large amounts of water through their body surface. This process is known as sweating and serves to regulate body temperature. If the body surface is 170 times larger relative to the body volume, then, in the first approximation, evaporation is also 170 times greater relative to the body's water reserves. Therefore, the miniature humans would have to drink a lot to avoid dehydration.

Yet, the miniature humans would hardly sweat at all. Their problem would be quite the opposite. There is another exchange process that occurs over the surface and also plays a significant role in technical chemistry. This is heat exchange. Chemical processes are constantly taking place inside the body, releasing heat. The purpose of these processes is to provide energy for all kinds of bodily functions, primarily muscle contraction, i.e., movement. The heat released in the process is initially a kind of loss. This energy cannot be used for muscle movement. However, the heat is not useless. It ensures that the body temperature remains permanently above the ambient temperature. This not only protects us from freezing into ice blocks in winter but also helps maintain bodily functions properly. Biochemistry operates best at a temperature of about 37 °C. The constant heat production inside the body is exactly balanced in the medium term by the constant heat dissipation through the body surface. The body temperature remains constant. If the body surface is increased 170-fold relative to the body volume, then the body also dissipates about 170 times more heat per body volume. To compensate, each cell would have to metabolize 170 times more carbohydrates or fats per unit of time. What initially sounds like a fantastic diet program would mean that the miniature humans would have to eat constantly. This is one of the reasons why, although there are very small animals, all animals below the size of a mouse are cold-blooded. This means they do not heat their bodies to a temperature of about 37 °C but adapt to the ambient temperature. Otherwise, the energy requirement would simply be too large. This relationship between body size and heat dissipation leads, among other things, to the fact that animals living in a polar climate are often significantly larger than their relatives from the tropics and is known as Bergmann's rule. It is one of the so-called ecogeographical rules that describe how the characteristics of closely related organisms differ in different habitats. A one-centimeter human with the same body proportions as a 1.70 m tall human would therefore constantly freeze. The episode does not report on this, but it can be assumed that Jadzia Dax has therefore set the temperature regulators on board the Rubicon to 35 or 36 °C.

However, miniature humans would have a significant advantage due to the high surface-to-volume ratio (which is not really of a chemical nature): Not only does the body surface increase in comparison to the volume. The cross-sectional area also increases in comparison to volume and mass. With respect to the cross-sectional area, their bones only have to bear one 170th of the weight. This is extremely helpful in falls. This can be observed in insects. If an ant were to fall from the tenth floor, it would not be seriously injured. On the one hand, it has to absorb hardly any body weight relative to the cross-sectional area. On the other hand, it does not fall very fast because it experiences a large air resistance relative to its body weight due to the high surface-to-volume ratio.[12] The same applies to the miniature humans. If the shrunken Miles O'Brien falls from a 1.5 cm high component in the circuit housing, it would be a significant fall relative to his body size, but he would not be seriously harmed (unless his fear comes true and he falls into a circuit). ◄

3.3 Tiny Atoms—Part 2

Also in the year 2374, an event occurs that raises further questions related to things smaller than atoms. Although there is hardly any time between the incident of the shrunken Rubicon and this event, there is a very great distance between them, even by Star Trek standards. This second event takes place at the other end of the galaxy. More precisely: in the Delta Quadrant. The crew of the stranded spaceship Voyager has to endure quite a lot on its seven-year journey home. The events of the 7th episode of the 4th VOY season, "*Scientific Method*", even cost a crew member his life. Quite apart from that, they raise some chemical questions again. What happened?

The Srivani are a technologically advanced species. To advance their research, they sometimes use highly questionable methods. This includes, among other things, conducting certain medical experiments. They do not conduct these tests on themselves but on the crew of the Voyager.

Now, there are ethical principles that apply to science. These are especially relevant for medical and biochemical research. This mainly concerns experiments on animals and humans. Experiments on living animals may only be conducted, among other things, if the corresponding experiment is unavoidable to find a cure for a disease, for example (animal testing for cosmetics is—at least within the EU—fundamentally prohibited, and even the import of such cosmetics is banned). The question of whether an experiment on living animals is scientifically justified is, of course, a question that every scientist must scrutinize closely. This is mandated by scientific ethics. But that is only the first step. If scientists themselves conclude that an experiment on vertebrates is unavoidable and ethically justified, the experiment must still be additionally approved by an external body. No scientist can approve such an experiment themselves. For experiments with humans,

---

[12] This should not be misunderstood as an invitation to throw small animals out of the window to test whether they really do not get injured.

which are sooner or later always necessary in the development of drugs, significantly stricter rules apply.[13]

In science, one repeatedly comes into contact with ethics (even if one is not a philosopher or theologian). An example of this is the so-called peer review. Scientific results are not simply published on a homepage, and if they are published in a scientific journal, the responsible editor does not simply decide alone whether the corresponding article will be published. If the editor of a reputable scientific journal[14] concludes that a contribution could be fundamentally suitable, they ask external scientists who are themselves active in this field and therefore potentially able to evaluate current research from this area. These scientists create reviews in which they give recommendations for acceptance, rejection, and possibly improvement of the contribution. The reviewers ("peer reviewers") are supposed to examine various aspects. The most important is, of course, the scientific methodology. That means: Is the approach suitable to answer the corresponding question, and was it conducted properly. Besides that, there are other questions of *good scientific practice* for the reviewers to examine. One question a reviewer should evaluate is whether they have ethical concerns regarding the corresponding research.[15]

Thus, scientific ethics has a high significance. At least for terrestrial science. The Srivani obviously see this much less strictly. Apparently, they are aware of the fact that their experiments can have extremely unpleasant consequences for the subjects. Therefore, they do not want the experiments to be conducted on

---

[13] When, which experiments are justified to what extent, is a difficult ethical question with very different views. These different views probably all have their justification. In Germany, there are legal regulations on this in the extremely lengthy paragraphs 7 to 10 of the Animal Welfare Act. But that does not absolve anyone from repeatedly asking the relevant ethical questions themselves. At this point, I am really glad that I research chemical energy storage and physicochemical fundamentals. But even if no experiments on living beings are involved, every scientist should always think about the ethical questions related to their research. Not only when researching living beings or atomic bombs, it is worth repeatedly questioning one's research in terms of its ethical consequences.

[14] The point "reputable" is unfortunately a big problem. There are a large number of so-called "predatory publishers" who publish anything for money. This is one of the reasons why it can play a significant role in science where something is published. With a well-known, reputable journal, one can assume that the review process has been properly conducted. Being able to classify the journal in this regard requires some experience among scientists and is often difficult for laypeople and yound students to understand.

[15] This question is, of course, very much aimed at experiments on living beings, but it is by no means limited to that. On the online platforms of some journals for submitting reviews, one even has to explicitly select whether one sees any ethical problems with the corresponding research (the system sometimes automatically asks this question to a reviewer, even if it is a theoretical work on energy storage).

themselves.[16] Therefore, they decide to conduct secret experiments on the visitors from the Alpha Quadrant.

They apparently anticipate that the crew of the Voyager would be less than enthusiastic about these experiments and would refuse consent even if asked politely. Accordingly, they simply come on board unasked and unnoticed. To do this, they use highly advanced cloaking technology that not only hides their two ships docked to the Voyager but also themselves. The phase-shifted Srivani can thus wander undisturbed through the ship and implant the craziest, also invisible, things into the crew. The phase shift used in this process raises some physical questions. One of the implants, in particular, also raises exciting chemical questions: a marker on the DNA.

The DNA, in which our genetic information is stored, is, simply put, a large molecule. Many atoms are lined up in two long chains, which are multiply connected and twisted into a double helix. Using a scanner she developed, B'Elanna Torres is now able to examine Chakotay's DNA. DNA examinations are no longer unusual even today. There are a whole range of biochemical methods to do this. However, the scanner developed by Torres is a microscope with capabilities that far exceed our current options. This scanner shows sharp images of the DNA, where you can even see the individual atoms. That's already quite good, but the truly remarkable part is yet to come. You can see even more. On individual atoms of the DNA, there are small markings that faintly resemble barcodes from the supermarket. Even viewers from the 24th century are astonished by such a level of submolecular technology. The Starfleet is far from this level. Ultimately, this is no wonder. The image of these tiny signs placed by the Srivani raises three questions from a chemist's perspective. First: How can it actually be seen? Second: How can it sit on the surface of an atom? Third: How can it exist at all?

Let's start with question number one. How can B'Elanna's scanner even produce an image of this marking? Quite obviously, this scanner does not create an image using visible light. In practice, the resolution of a microscope is often limited by factors such as the quality of the optical lenses. However, if you push the technology to its limits, you eventually reach another boundary for resolution: the wavelength of light. Simply put, you cannot resolve structures that are smaller than the wavelength of the light used.

For light visible to us humans, the wavelength is approximately between 380 and 750 nm. The short-wavelength light at 380 nm (or slightly more) is violet. The long-wavelength light at 750 nm (and slightly less) is red. The entire spectrum of

---

[16] At this point, one can, of course, ask whether the human decision to conduct corresponding experiments on other species does not have parallels to the behavior of the Srivani. Quite obviously, the Srivani not only practice the corresponding experiments but also see no reasons to restrict them in any way for ethical reasons.

the rainbow spans between these two extremes.[17] If you want to take a picture of a DNA molecule with visible light, it would be difficult to even recognize the individual atoms. A carbon atom has a diameter of only about 140 picometers. Most of the other atoms in the DNA are even slightly smaller. Only the few phosphorus molecules are slightly larger at 200 picometers. Comparing these sizes, it becomes apparent that even the largest atoms are smaller than the wavelength of the shortest visible light. The difference becomes especially glaring when you not only look at the numerical values but also at the unit. The unit of wavelength is nanometers. That is one billionth of a meter. The unit of atomic diameters is picometers. That is one trillionth of a meter. The atoms are therefore more than a factor of 1000 too small to be seen with visible light—not to mention resolving any even smaller structures on them.

With shorter-wavelength light, which is no longer directly visible to the human eye, the lower limit of resolution can be shifted downward. If the wavelength is only slightly reduced, it is referred to as ultraviolet light. If the wavelength is further reduced, one eventually enters the range of X-rays. With very hard X-rays, one indeed slowly enters a range where resolutions on the order of individual atoms would theoretically be conceivable. Chemists are already using such techniques today. For example, X-ray diffraction (XRD) is used to determine the arrangement of atoms in crystals. X-ray photoelectron spectroscopy (XPS) is used to analyze the chemical composition of surface layers that are only a few atoms thick. Such techniques are used, among other things, in the investigation of catalysts. To achieve microscopy with even higher resolution, one leaves the realm of electromagnetic waves, to which light and X-rays belong. Instead, one can use electron beams. However, it should not be forgotten that even electrons have a wavelength. With very high-energy electrons, this wavelength is quite small, which is why resolutions of about 100 picometers are challenging but fundamentally possible. The individual atom is thus slowly coming into the visible range for today's chemists. Structures smaller than an atom, however, are still far outside the realistic range for the foreseeable future. Besides the technical implementation, which has not yet been achieved, there is currently not even a type of radiation known that could resolve something so small. If we are slowly approaching the resolution of individual atoms at the beginning of the 21st century, it does not

---

[17] The spectrum of light has great significance in chemistry. Even in the first semester (and hopefully earlier in school chemistry classes), every chemistry student must conduct practical experiments on flame coloration. For example, if you hold a sodium salt in a flame, the color of the flame changes due to the characteristic spectral lines of sodium and becomes yellow. With a lithium salt, it would turn red, and with a potassium salt, violet. Each chemical element has its own characteristic spectral lines by which it can be optically identified. Individual elements were even discovered this way. One of them was discovered in 1868, for example, when an unexplained bright yellow spectral line was found during the examination of the sun's chromosphere during a solar eclipse. Since this line could not be assigned to any known element, it was concluded that it came from a previously unknown element. This element was eventually named helium, after the Greek word for the sun.

seem entirely far-fetched for the middle of the 24th century. Even if we currently cannot say how.

The second question is more difficult: Can a tiny, submolecular structure be simply placed on the spherical surface of an atom? Even B'Elanna Torres finds it a mystery with which tool this should be done. But let's assume that such a tool exists. The question remains: Is there even a spherical surface of the atom on which the structure can be placed?

According to the classical atomic model, there is indeed a solid surface of the atom. The term atom itself implies this. The Greek word "atomos" means indivisible. The idea was that the atom is the smallest unit and that it does not consist of other, smaller components. Modern chemistry, however, abandoned this idea at the beginning of the 20th century. Today we know that atoms consist of two parts: the atomic nucleus and the electrons.[18] The internal structure of an atom is somewhat complicated, and it becomes even more complicated when considering atoms that are bound into molecules. A very popular and simple representation of this is the so-called Bohr model of the atom, which we already touched upon when discussing hydrogen. It was developed by Niels Bohr in 1913 and is fundamentally well-suited to appeal to a Star Trek fan. The Bohr model of the atom assumes that atoms are structured like small solar systems. In the center is a large, heavy star (the atomic nucleus), and around this central star, one or more small planets (the electrons) orbit.[19] Let's imagine the atom in this way as a first step and also think of the markings applied by the Srivani. Where should these be placed? The atom is not a sphere with a surface. It is a solar system with (in the case of carbon atoms) six planets called electrons. Accordingly, one cannot simply place something on a non-existent outer surface.

The Bohr model of the atom was an important step towards understanding atomic structure. However, it is far from the final word on the matter. In reality, electrons are not solid particles that orbit the atomic nucleus. Just like light, electrons have a dual nature: they are both wave and particle.

The dual nature of light often leads to confusion because people ask whether light is a wave or a particle. The answer is truly: both at the same time.

The wave nature of light is evident, for example, in light scattering. When observing water waves hitting an obstacle, one notices that they seem to bend around the obstacle. Another example is the superposition of waves, known as interference. Depending on how two waves overlap, they can either amplify or weaken each other. What can be observed on a water surface can also be observed

---

[18] Strictly speaking, the atomic nucleus (apart from the simplest hydrogen isotope) consists of several smaller particles; primarily protons and neutrons, which are held together by the mediation of other particles. For chemistry, the division into atomic nucleus and electrons is usually sufficient, as chemical bonds are mediated by the electrons.

[19] What the Trekkie might miss in this model to be completely happy are moons orbiting the electrons. Then it would be truly perfect.

with coherent light. The Michelson-Morley experiment, the fundamental discovery that led to the theory of special relativity, is based precisely on this.

At the same time, light is transmitted in the form of particles, called photons. Ultimately, photovoltaics or the biochemical conversion of carbon dioxide and water into oxygen and sugar in photosynthesis is based on this. Both natures are always present. However, there is a tendency for the particle nature to come to the forefront when the light is very short-wavelength, that is, very high-energy. Conversely, the wave nature is more pronounced when the light is rather long-wavelength.

However, this dual nature is not limited to light. All particles are simultaneously waves. The more energetic (practically speaking, usually: the more massive) they are, the less the wave nature emerges. For heavy particles, the wavelength is significantly smaller than the particle itself. Thus, the notion that the atomic nucleus is a particle and not a wave is generally already a very good assumption. In contrast, the wave nature is much more pronounced in the much lighter electrons.

Instead of imagining electrons as particles that orbit the center like planets, it is better to envision them as a cloud. The location of the electron is not really clearly defined, but there is only an area called an orbital within which the electron resides. As mentioned, it does not have a clearly defined location, but only a probability of being at individual points within the orbital. Since each orbital can accommodate a maximum of two electrons, all atoms, except for hydrogen and helium, have multiple orbitals. These additional orbitals are then no longer spherical but take on complex shapes to fill the space around the atomic nucleus as efficiently as possible. In practice, however, the atom is often simply imagined as a sphere with a clearly defined surface. For many questions, this is a reasonable and quite practical approximation of reality. However, an atom does not have a solid spherical surface on which a mark could be placed.

Some More Details

Upon closer examination of the image of the DNA recorded by B'Elanna's scanner, one notices that the surface of the spherical atom is not uniform but has a certain shading. Here, the creators of Star Trek have once again proven to be very precise, as the distribution of electrons described only by a probability density also undergoes fluctuations (in the Bohr model of the atom, this can be simplified by imagining that the electrons orbit the nucleus and sometimes there are more electrons on one side than on the other). Since the electrons are negatively charged and the atomic nucleus is positively charged, this leads to an unequal distribution of charge within the atom, resulting in a so-called spontaneous dipole. Overall, the particle is still electrically neutral, but it temporarily has a positive and a negative end. This can, in turn, induce a dipole in a neighboring atom or molecule (meaning: electrons are shifted within the molecule, so that this molecule also temporarily becomes a dipole). Since the positive side of the spontaneous dipole is closer to the negative side of the induced dipole than to its positive side (or vice versa), the two molecules attract each other. This effect is the cause of the so-called van der Waals forces between molecules. ◄

Finally, there remains a third, already hinted-at question raised by the submolecular marking of the Srivani: Can such a small structure even exist?

Every known structure is made up of chemical elements and compounds. However, they all consist of atoms. Printing a barcode on an atom, therefore, requires an ink that is not made of atoms. Such a thing does not exist. Neither does it exist today, nor does it seem to be known to B'Elanna Torres in the 24th century. The Srivani are obviously much further advanced. How exactly they manage this, we cannot say, but is it fundamentally unthinkable?

In chemistry, one usually deals with no more than three elementary particles: electrons, protons, and neutrons. In the practical work of a chemist, it is almost always only the first two.[20] However, there are many more elementary particles in the universe. Hundreds have already been discovered today, and the names of many of these elementary particles are certainly familiar to Star Trek viewers. Although no one today would have any idea what a chemical compound based on other elementary particles might look like.[21] But that does not mean it could not be possible. In principle, it would be conceivable to build structures from other elementary particles that are much smaller than the elementary particles that make up the matter surrounding us and of which we ourselves are composed. However, if one were to attempt to build structures significantly smaller than atoms, one would run into trouble with a gentleman named Werner Heisenberg.

The Heisenberg uncertainty principle, formulated in 1927, states that the position and velocity of a particle cannot be determined with arbitrary precision. In practice, this limit of accuracy is mostly completely irrelevant because the practically achievable measurement accuracy is significantly worse. However, if one were to push the measurement technology to its limits and perform the best possible measurement, one can still never measure more accurately than the Heisenberg

---

[20] Electrons are enormously important in chemistry because they ultimately mediate the bonding between atoms to form molecules. Additionally, they are transferred from one atom to another in a very important class of chemical reactions, the redox reactions. Protons are also transferred in a specific type of reaction. These are the so-called protolysis or acid-base reactions. In this process, no elementary particle is removed from one atomic nucleus and transferred to another nucleus. Rather, it is simply a positively charged hydrogen atom (or better, ion). The atomic nucleus of the most important hydrogen isotope consists of just a single proton, and if you take away its only electron, only the proton remains as the hydrogen ion. In protolysis, these protons are transferred from an acid molecule to a base molecule.

[21] Muonium could be understood as a kind of atom made of other elementary particles. In this structure, discovered in 1960 by a team led by Vernon W. Hughes, an electron orbits an antimuon (as in conventional atoms). Muonium thus has a certain similarity to a hydrogen atom. However, its mass is significantly smaller. The same applies to its lifespan, which is in the range of microseconds.

uncertainty principle allows. In fact, the Heisenberg uncertainty principle does not say that one cannot determine position or velocity with arbitrary precision. However, if one does, the other quantity becomes arbitrarily imprecise.[22]

If we now use miniature atoms, consisting of whatever elementary particles, to build such small microstructures as the Srivani do, we would have to determine the position of the individual atoms incredibly precisely. To form something like chemical bonds, each miniature atom would need to have a position relative to the other miniature atoms that is enormously precise; much more precise than in normal molecules. Consequently, the velocity of the particles would become very imprecise. How one is supposed to form molecules with such an indeterminate velocity into a solid structure is a completely unresolved question.[23]

However, it can be assumed that there is a solution for this in Star Trek. In the 12th episode of the 6th TNG season, "*Ship in a Bottle*", we learn, for example, that the transporters have a Heisenberg compensator. The Heisenberg uncertainty principle would actually make beaming impossible. Fortunately, the engineers in Star Trek have found a solution for this. Unfortunately, we do not learn how the Heisenberg compensator works.

---

[22] Strictly speaking, the Heisenberg uncertainty principle is not about an uncertainty in velocity, but in momentum. Momentum is the product of mass and velocity. If we assume that the mass of a particle is fixed, then it is ultimately the velocity whose accuracy is coupled with the position by the Heisenberg uncertainty principle. The limit in accuracy results from the product of the accuracy of position and momentum. This product cannot be greater than Planck's constant. Since this constant, with a value of about 0.000000000000000000000000000000000066 J·s, is very small, the effect is hardly noticeable in practice. In our case, however, the Heisenberg uncertainty principle causes serious problems.

[23] The same effect would most likely have caused difficulties for the crew of the shrunken Rubicon from our previous example. The 1 cm tall humans would not have been able to observe it visibly, but with the shrunken atoms, their biochemistry would probably have already started to behave significantly differently than it should.

Chemistry and Its Speed

4.1 The Salt Vampire of M-113

The 5th episode of the 1st TOS season, *"The Man Trap"*, was one of the first episodes of Star Trek ever. In it, Dr. McCoy meets his old flame Nancy Crater again on the planet M-113. As the episode progresses, it turns out that she is not only married but has been dead for quite some time. The alien creature that consumed her, however, is capable of taking any form and is now posing as Nancy. Although! It didn't really consume her. It only needed one component of her body: salt.

This life form has many suction cups on its hands. With the help of these, the creature seems to be able to suck all the salt out of its victims' bodies. For this salt vampire, as it is fondly called by Star Trek fans, table salt (chemically: sodium chloride; NaCl) is apparently of central importance. This is a statement that initially applies equally to the human organism. We cannot live without salt. On the other hand, the daily requirement of a human for table salt is only a few grams. Today, unlike in the past, salt is very easily available. However, our bodies are still conditioned to get as much of this vital and sometimes scarce mineral in nature as possible. As a result, we modern humans in Europe mainly have the problem that we tend to eat too much salt. This has some significant medical consequences.

Of course, our biochemistry doesn't work without salt either. If a salt vampire from the planet M-113 were to suck all the sodium chloride out of our bodies within seconds, it would indeed be fatal. And also within seconds, as some crew members of the Enterprise have to experience firsthand. The reason for this is simple. Salt has a number of vital functions in the body. Removing all salt from the nerves would kill us instantly. It plays a role in the transmission of nerve impulses. Without nerve impulses, there is no muscle contraction. Without muscle contraction, there is no heartbeat. Without a heartbeat, there is no oxygen supply to the brain and other organs. This would mean the death of the victim in a very short time.

The need for salt is significantly greater in the salt vampires of M-113 than in us humans.[1] What exactly they need the salt for is unfortunately not clarified. From a chemical point of view, however, it is clear that the consumption of salt cannot really serve as an energy source. Salt may look similar to sugar. Nevertheless, it can only very limitedly take over its function as an energy carrier. Salt cannot be oxidized to gain energy for any body functions. There is also no chemical reaction with table salt as a starting material that would be suitable as an energy source for living beings. But would it really be unthinkable for alien life forms to use salt for energy provision?

There is indeed a type of power plant that is based on salt: the so-called osmotic power plant. For this, you need water sources with different salt concentrations. In practice, you find this, for example, at the mouth of rivers into the sea. There, large amounts of both saltwater and freshwater are available. In an osmotic power plant, as the name suggests, an effect called osmosis is used. We will get to know osmosis in more detail later. In short, osmotic power plants use the fact that freshwater passes through certain membranes even when there is saltwater with a significantly higher pressure on the other side. This allows an even higher pressure to be built up in the saltwater, which can be used in a turbine to generate power. This is possible. However, this technology has not yet progressed beyond the stage of individual demonstration plants. In the foreseeable future, it is unlikely that osmotic power plants will play a significant role in our energy supply. Nevertheless, it is possible to generate energy from salt. At least as long as you can get a solution with a low salt concentration from somewhere. As we will see, this will be somewhat difficult for a salt vampire. But who knows what their organism actually needs the salt for so urgently? Maybe it has a completely different function for their biochemistry.

The much more exciting question, however, is how the salt vampire sucks the salt out of its victims. As mentioned, it has several suction cups on its hands. However, it is not the case that the bodies of humans or other living beings have taps where it could dock with its suction cups to suck out the salt. A "normal vampire" has it much easier in this regard. Although there are no taps for blood on the human body either, it is relatively easy to create one. As we know from countless vampire books and movies, the vampire only needs to pierce its victim near an artery with its sharp teeth. In a pinch, it could even pierce a vein, of which there are several near the surface of the body. Then it would have to suck a bit harder

---

[1] The salt requirement seems to be so high that the species eventually became extinct because all the salt on M-113 was used up at some point. It can't really be used up. No matter what chemical reaction you do with salt: the sodium and chlorine atoms from which it is built remain as such, and the total amount on the planet does not change. Nevertheless, it is conceivable that at some point all the table salt is "used up" in the sense that we "use up" water. Water is not destroyed when washing and drinking. It is still water. However, it becomes contaminated, meaning it is dirtier after use than before. It is "used up" in the sense that it is still there but can no longer be used immediately. Something similar could have happened with the salt on M-113.

itself, as the blood pressure is lower there. In principle, however, it is still a fairly simple process. The situation looks much more complicated for a salt vampire.

The problem is the distribution of the salt. There is no salt chamber in the human body, even if the salt concentration may slightly differ in the various organs. By and large, the salt is quite evenly distributed in the human body. This applies not only to humans but ultimately to all higher living beings. So, it is not enough for the salt vampire to apply its suction cups in the right place, analogous to the teeth of its bloodsucking namesake, because there simply is no right place. No matter where it starts to suck its victim: It would not simply suck out the salt, but every liquid or dissolved chemical substance.

Let's take a closer look at the salt vampires from M-113. As mentioned, there is a kind of suction cup on their hands. The function of a suction cup is normally not to suck out, but to suck on. Octopuses and other animals have their suction cups not to suck out prey animals, but to suck on. On the one hand, the eponymous effect can serve this purpose. By creating a vacuum at the suction cups, the animal holds onto a surface (this surface can then again be the body of a prey animal). Besides the vacuum effect, rather chemical effects can also come into play with suction cups. Some animals secrete glandular secretions to hold on. Such a secretion can not only serve the (very important) sealing of a vacuum. It can also contribute to adhesion itself. This effect is called adhesion and is a significant working mechanism for adhesives. There are attractive forces between molecules. However, the strength of these attractive forces can vary greatly. Many factors come into play. Among other things, the size of the molecules and their chemical structure play a significant role. Simply put, an animal must apply a glandular secretion over its suction cups in such a way that it maximizes the contact between the surface and the suction cup. On the other hand, the molecules of the secretion must have strong attractive forces both to the surface and to the body of the animal.

However, the salt vampire does not suck onto its victim at all. While it extracts the salt from them, the victims seem to be almost paralyzed. This suggests that it actually secretes a substance that chemically incapacitates the victims. However, this cannot explain the actual function of salt sucking. For this, it needs something that allows it to absorb only the salt. The other components of its prey's body must remain where they are. Accordingly, the beings from M-113 do not need openings (like a mouth), but membranes. Only in this way can salt be absorbed and everything else be retained by the membrane. Thus, the mystery seems to be solved. The suction cups represent membranes through which the salt vampire selectively[2] sucks the salt out of its prey's body.

---

[2] Selective is a term that appears more often in chemistry. Selectivity means that only one of several possible chemical reactions occurs or, as in the case of the membrane, only the molecules of a specific substance are allowed through and all others are retained. In practice, there are rarely selectivities of 100%. Instead, one often has to deal with side reactions or other substances passing through the membrane. Therefore, the goal in practice is often to achieve the highest possible selectivity (a lot of the desired substance, little of the undesired).

Even if that is the answer, it raises three new questions at once:

1. How can a membrane let salt through and retain everything else?
2. How does the salt vampire get the salt to pass through the membrane at all?
3. How can it extract the salt from its victims so quickly?

Let's start with the first question. We need a membrane that allows salt to be separated from water. Blood or lymph fluid contains the desired salt along with a lot of water (in addition, there are many other things, but these are all larger molecules or even whole particles that can be easily separated). Membranes that separate water and salt do exist. In technology, they are used, for example, in seawater desalination. The process is called reverse osmosis. As the name already suggests, the principle is derived from another effect that is "reversed" in technology: The biological effect of osmosis. In osmosis (just like in reverse osmosis), there are selective membranes. Such a selective membrane lets one substance through (the chemist says: it is *permeable* to it). It does not let the other substance through (for it, it is *impermeable*).

The cell membranes that enclose not only the cells of us humans but the cells of all living beings are such semipermeable membranes. The Latin prefix "semi-" means half. Half is let through, the other half is not. Whether it is really exactly half is another matter. What is crucial is something else. The membrane is permeable to water and impermeable to salt. In other words: Exactly the wrong substance is let through. The salt that the salt vampire actually wants to let through does not pass through.

Therefore, it would need a completely different type of membrane. A membrane that is permeable to salt but impermeable to water. At this point, the salt vampire would need a membrane in its suction cups that is significantly different from what known biology offers. This is not entirely trivial. Compared to the further challenges, however, it is rather the smallest problem. The second question proves to be considerably more critical: How does one get the salt to pass through the membrane at all?

Let us assume for a moment that the salt vampire possessed a hypothetical, perfect membrane. This membrane would not only be one hundred percent selective. That is, it would only allow the desired salt to pass through, while not a single water molecule or other substance would manage to get through. At the same time, the diffusion resistance for salt would also be negligibly small. With membranes, there is often the problem that the desired substance does pass through, but it is slowed down by the membrane. The unwanted substances are merely slowed down even more. This is how the separation effect of the membrane comes about. Let us now simply imagine that our perfect membrane poses no obstacle to salt at all. We assume that the salt can move through the membrane as if it were not there at all, but the membrane would magically only hinder the water. What would salt in an aqueous solution do at this membrane?

Let us imagine such a membrane. On one side, there is salt water, and on the other side of the membrane, there is also salt water. Both the sodium and chloride ions that form the salt can pass through the membrane unhindered. The water, however, cannot. Let us say that on the left side of the membrane is Captain Kirk's blood. On the right side is the blood of the salt vampire, who is trying to extract all the salt from the Captain. There would be nothing to stop the sodium and chloride ions from migrating through the membrane to the salt vampire. Since the water would be held back, the poor Captain would lose more and more salt over time. At least if there were not the salt on the other side. There is also salt in the salt vampire's blood. Even if there had been no salt in his blood at the beginning, his blood would also contain salt as soon as he had extracted a significant amount. This salt on the right side could also migrate through the membrane and replenish Kirk's body's salt supply. Ultimately, what matters is the net salt exchange. This means nothing other than the amount of salt that migrates from left to right minus the amount of salt that migrates from right to left.

Let us put ourselves on the molecular level. This means in this context the order of magnitude of the ions of the salt. To migrate through the membrane, an ion must first hit the membrane. The more ions hit the membrane per unit of time, the more migrate through the membrane. In the end, there is a net migration of ions from the side where many ions hit the membrane to the side where few ions hit the membrane per unit of time. The number of ions hitting the membrane per unit of time depends on two factors: temperature and concentration.

The molecules and ions in saltwater are not rigidly arranged so that each has its fixed place, but they move more or less freely. The higher the temperature, the faster they move. As a result, more ions hit the same area per unit of time. However, different temperatures on both sides of the membrane are not a really practical method to drive a substance through a membrane. On the one hand, the temperatures on both sides of a thin membrane equalize very quickly. On the other hand, the body temperatures of humans and salt vampires do not seem to differ significantly. Much more important is the effect of concentration. When the concentration is high, many ions are present in the same volume, and thus many ions hit a given membrane area per unit of time. If there is a high salt concentration on the left side and a low one on the right side, there is a net migration of the salt ions from left to right. Ions also migrate from right to left, but many more ions migrate from left to right. This is the net migration called diffusion. In summary, the rule is: A substance diffuses from an area of high concentration to an area of low concentration. This ultimately applies regardless of whether there is a membrane in between or not. The membrane can only inhibit diffusion, not cause it.

This has a very crucial consequence for the salt vampire: He must have a lower salt concentration in his blood than in the blood of his victim. At first, this does not sound so difficult. At the beginning, it is not. However, if he extracts more and more salt from his victim, the salt concentration on both sides changes. On the left side (in Captain Kirk's body), it decreases. On the right side (in the salt vampire), it increases. If the salt vampire really wanted to extract all the salt from his victim, he would have to reduce the concentration in his body to zero. For the salt ions to still migrate through the membrane to the salt vampire, he would have to keep the salt concentration in his blood at zero as well (strictly speaking, even below zero, which, however, makes no chemical sense). Since salt is constantly flowing into his body from his victim's body, this is somewhat difficult. Since his body is obviously not infinitely large, the salt concentration will inevitably increase. The extraction of salt will therefore come to a halt as soon as the concentrations on both sides are equal. If we assume that salt vampires are about the same size as an average human and at the beginning of the extraction process there was no salt in his body (which explains why he urgently needs some), then he could at most steal half of his victim's salt. Once he has extracted half, the salt concentrations would be balanced due to the approximately equal body volume. The net diffusion would come to a halt, and the victim would have a massive salt deficiency. However, not all the salt would be extracted, but only half.[3]

So, is the salt vampire doomed to a cruel death from salt deficiency because his process of salt absorption from human victims does not work properly at all? Well, it is indeed difficult. However, there is still one possibility: electrodialysis.

Salts consist of ions. In the case of table salt, these are sodium ions and chloride ions. The sodium ions are positively charged cations, while the chloride ions are negatively charged anions. When an electric field is applied, the sodium ions move towards the negative pole and the chloride ions towards the positive pole.[4] A sufficiently strong electric field can indeed cause the ions to move against the concentration gradient; in other words, from an area of low concentration to an area of high concentration.

For a salt vampire, however, it is not enough to simply generate an electric field and thereby pull the ions of the salt towards itself. If it charges its suction cups positively, for example, it attracts the anions (i.e., the chloride). The cations (the sodium), on the other hand, would be repelled. This would not only cause it to miss out on half of the ions. In no time, its positive charge would be gone, as the attracted negative chloride ions would compensate for the positive charge. Conversely, the same would apply if it charged its suction cups negatively to attract the sodium ions. This is where electrodialysis comes into play. This process uses a combination of membranes and an electric field. The method is quite common in technical chemistry. It is not yet known in biological organisms, but who knows what nature on M-113 might produce. Let's take a look at the process (Fig. 4.1).

First, there would need to be a membrane on the outside of the suction cups that allows both ions to pass through. Inside the suction cups, the salt vampire would need to generate an electric field. Technically speaking, we would

---

[3] A little more could theoretically be possible if the salt vampire's blood had a significantly higher solubility for salt than that of humans. However, since water is already one of the best solvents for table salt, it would be chemically difficult to find a suitable substance.

[4] The negative pole is also called the cathode, from which the term cations comes, because they migrate there. The positive pole is called the anode, from which the term anion is derived analogously.

4.1 The Salt Vampire of M-113

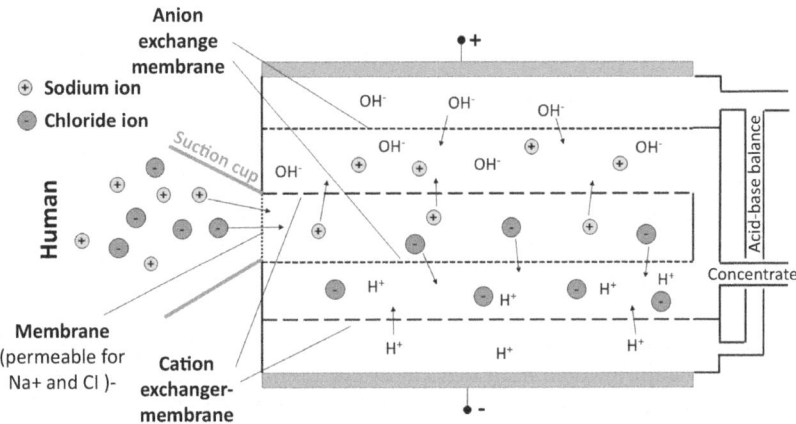

Fig. 4.1 Schematic representation of an electrodialysis with which the salt vampire could selectively extract only the salt from its victims

be dealing with an electric capacitor. Within this capacitor, there are two different types of membranes responsible for the chemistry. The membranes differ in what they allow to pass through. On the one hand, there are anion exchange membranes. As the name suggests, they only allow anions to pass through. Unsurprisingly, the second type is cation exchange membranes. They only allow cations to pass through. These membranes are arranged perpendicular to the electric field (i.e., parallel to the plates of the capacitor). When the negatively charged chloride ions move towards the positive pole, they encounter an anion exchange membrane. Since they are anions themselves, they pass through without any problems. Conversely, the positively charged sodium ions eventually encounter a cation exchange membrane during their migration. As cations, they also pass through largely unimpeded. However, this is not the end of the process. We would now have separated sodium and chloride ions. This would lead to strong accumulations of electric charge. These charges need to be balanced. For this, the salt vampire would need to use other ions. The obvious solution would be positive hydrogen ions H^+ and negative hydroxide ions OH^-.[5] These two ions are always present in water. Through the so-called autoprotolysis, water molecules continuously split into these two ions, which then recombine into water molecules. Through another anion exchange membrane, hydroxide ions can reach the sodium ions. Through a

---

[5] Strictly speaking, there are no individual hydrogen ions H^+ swimming around in the water. Instead, the hydrogen ions combine with water molecules to form oxonium ions H_3O^+. For simplicity, however, we will pretend here that they are simple hydrogen ions. This makes no difference for understanding the basic principle of electrodialysis.

cation exchange membrane, hydrogen ions can reach the chloride ions. This balances the electric charges again.

The last problem remaining is the pH value. The pH value indicates whether an aqueous solution is acidic (then it is less than 7) or alkaline (then it is greater than 7). Roughly speaking, it is a measure of the concentration of hydrogen and hydroxide ions. If a large number of hydrogen ions accumulate in an area (or hydroxide ions are removed), an acid is present. If many hydroxide ions accumulate (or hydrogen ions are removed), a base is present. Both would be disadvantageous in the long run. Therefore, it is important that the acid and base are each neutralized. This can be achieved by mixing them. When hydrogen and hydroxide ions meet, they neutralize each other and form a water molecule. This is exactly what the salt vampire would need to do in its electrodialysis suction cups.

Overall, this would require two separate blood circulations. In one, the salt accumulates. In the other, acid (hydrogen ions) and base (hydroxide ions) are continuously exchanged and thus neutralized.

Extracting all the salt from a human in this way and transferring it to the body of another living being would certainly be challenging. The question alone of how a biological organism is supposed to generate the corresponding electric field would be anything but trivial. The correspondingly intricate membranes, permeated by different blood systems, would also be complex. When you look at the variety of living beings in nature, at least this part does not seem so far-fetched. The salt vampire could therefore fundamentally be a reality.

Finally, we should also take another look at the third question. How can he extract the salt from his victim so quickly? His victims are all dead and completely saltless within a few seconds. If Dr. McCoy had not used a phaser quickly enough to stop him, Captain Kirk would have shared the fate of several crew members in no time. However, the speed at which salt can travel through a membrane is limited.

The larger the membrane surface area, the faster it goes. However, this area is rather manageable because he only absorbs the salt through his palms. Even if the salt absorption through the membrane were fast enough, the question would still remain how the salt gets there so quickly. In the human body, the salt is, as mentioned, more or less evenly distributed. Even if everything were in the blood (which is by no means the case), it would certainly take a few minutes for all the blood to pass by the salt vampire's hand once. Moreover, most of the salt is not in the blood but within the cells of the body. These are surrounded by cell membranes, which are nothing other than osmosis membranes. They let water through. Salt, on the other hand, does not come out so easily. Consequently, most of the salt is retained by the membranes in the cells and does not immediately enter the blood. Therefore, it would take a very long time to extract all the salt from a human.

Although we have found a possible explanation for the basic functionality of salt extraction, the speed remains one of the mysteries that still need to be solved in the infinite expanses of space.

Excursus

The Cloud-Shaped Iron Vampire

The salt vampire from *"The Man Trap"* is not the only extraterrestrial being that extracts an essential substance from humans. Another example is the cloud-like creature based on dikironium[6] that the Enterprise encounters some time later in the 18th episode of the 2nd TOS season, *"Obsession."* This bodiless creature feeds on the hemoglobin of its victims. Hemoglobin is a very complex chemical compound. At its center is an iron ion, around which four complex proteins are arranged. Hemoglobin is vital because oxygen binds to it. Thus, it helps transport the essential oxygen from the lungs to the other organs. Without hemoglobin, we would die within a very short time, as some crew members have to experience for themselves.

Extracting hemoglobin from human blood is significantly more challenging than extracting salt. While the ions of salt are dissolved in the blood and the water inside the cells, hemoglobin is part of the red blood cells (even though these themselves consist of about 90% hemoglobin). Therefore, it is not a free molecule that can be individually extracted. First, bonds must be broken to release hemoglobin. The dikironium cloud creature could inject an appropriate enzyme into the blood of its victims for this purpose. Subsequently, the hemoglobin must be extracted from the body in some way. Doing this selectively is significantly more challenging because it is not dealing with small single-atom ions but with a large, highly complex molecule. Selectively detaching the hemoglobin from the red blood cells and then extracting only the hemoglobin while leaving all other proteins in the body would be a true masterpiece of cosmic biology. However, the dikironium cloud creature is a much more highly developed life form than the salt vampire or us humans. ◄

4.2 A Thirsty Virus

In the 25th episode of the 2nd TOS season, *"The Omega Glory,"* an away team from the Enterprise makes a strange discovery. The group had beamed aboard the Starfleet spaceship USS Exeter. However, they find no one there. The only thing they find are the crew's uniforms. These are simply scattered around the ship. Empty, without the people who belonged in them. Well, almost empty. The away team finds a few crumbs in them. Small crystals that are all that remains of the crew.

As it turns out later, everyone on board is dead. Killed by a virus. What is truly remarkable, however, is what this virus did to the people. It caused all the water to

---

[6] The chemical nature of dikironium would also be an exciting question. Unfortunately, this substance will probably only be discovered in the distant future.

be extracted from their bodies. All that remained were the crystals. These contain everything that was not water in the human body.

What happens aboard the Exeter is not only an extremely threatening situation. After all, the away team from the Enterprise faces the same fate. It is also a highly fascinating process.

The whole thing raises a number of questions. The removal of water from the body of a multicellular organism by a virus is remarkable. First of all, the question arises of how the virus does this. And secondly, one wonders why it does it at all. To understand this, we first need to consider what viruses actually are and what they do.

Viruses are microscopically small... Well, what exactly? In biology, we deal with organisms. But viruses are not organisms. What characterizes an organism is that it has a metabolism. This means that it converts chemical substances. For example, organisms take up sugar and convert it with oxygen so that they can use the released energy. Or they link amino acids at their ribosomes to form proteins, which are then used to build the structures of the cell. A virus, on the other hand, does none of this. When you look at it more closely, you find that it is not even a proper cell. Practically everything that makes up a cell is missing. There are no ribosomes, no mitochondria. Essentially, there is only the genetic information and a shell that surrounds it. That's all.

Therefore, a virus cannot reproduce on its own. After all, it does not have a metabolism. Reproduction means that one "organism" becomes two (or more). For this, everything must be duplicated. In the case of the virus, the nucleic acid with the genetic information must be duplicated. Likewise, the protein shell must be duplicated. Both would be a chemical reaction. Exactly that is something a virus cannot do. It does not have a metabolism. It cannot provide the necessary chemical building blocks for the new nucleic acid and the new proteins at all. And even if they were available, it still could not do anything with them. For that, it would need its own metabolism. To reproduce nonetheless, viruses use a completely different strategy. They use organisms and make them do this work for them.

The envelope of viruses is such that host cells take up the viruses upon contact. The virus essentially tricks the cell into thinking it is something worth taking in. Instead of valuable nutrients or the like, the cell only gets the virus's nucleic acid. This viral genetic information is now inside the cell. Some cunning viruses do not start right away. Instead of reproducing immediately, they first integrate their genetic information into that of their host. Sooner or later, however, all viruses begin to reproduce. But they do not do this themselves. Their nucleic acids with the viral genetic information are replicated inside the cell. The same chemical process that is supposed to duplicate the cell's genetic material during cell division now multiplies the virus's genetic information. This nucleic acid does something else as well. Normally, nucleic acids like DNA or RNA serve to form proteins with the information stored on them. For this, amino acids assemble at the ribosomes as specified by the nucleic acid. But now it is no longer the cell's nucleic acid, but that of the virus. The ribosomes thus begin to produce the proteins for viral envelopes instead of things that would be useful for the cell. This consumes the cell's valuable resources. When enough viral nucleic acids and proteins for viral envelopes have been formed, they exit the cell again. In the worst case for the cell, it bursts in the process. In the best case, the viruses exit through the cell membrane without killing the cell. The cell then just has to keep producing new viruses and using its own resources for the virus.

This description of the mechanisms of viruses is undoubtedly greatly simplified. However, it is sufficient to understand what happened to the crew of the Exeter and where the problems lie. Before we turn to the question of how the alien virus does this, let us first ask why it does it. Why does it drain all the water from its host's body? Since it does not have a metabolism, it cannot absorb the water itself. Rather, it must make its host cells release all the water into the environment. This kills its own host and dries out the cells in which it is replicated. So why would it make the cells release water into the environment?

The explanation could be the same as why one has to cough or sneeze when having a cold: to spread the virus. A virus is not only interested in having its nucleic acid and protein shell replicated by the host cells. It also wants to spread. Other hosts need to be infected; otherwise, its reproduction remains very limited. Therefore, many viruses trigger processes that serve their spread. Coughing or sneezing are examples of this.

If a cold virus succeeds in making its host sneeze, the host exhales very quickly in the process. Small water droplets are carried along. An aerosol is formed. Many small droplets float in the air. These small droplets contain viruses. If someone else inhales these droplets, they can become infected themselves. The viruses can now use their body for reproduction as well. Perhaps the viruses on board the USS Exeter do something similar. They might (through some much more sophisticated mechanism than sneezing or coughing) distribute the water in their host's body into the air. Not as vapor, because then the water molecules would float through the air individually. A virus cannot cling to a single water molecule to be transported. After all, it is many times larger than a water molecule. Instead, it is more effective for a virus to form a mist. This consists of many small droplets. And viruses could be present in these droplets. This would explain why the entire crew of the Exeter became infected so quickly. If the entire air is indeed contaminated with virus-laden aerosol droplets, it is difficult to defend against it.

The fact that it is not foggy on board the Exeter does not necessarily mean that the water from the crew's bodies did not form a mist. After all, some time passes before the Enterprise arrives. In the meantime, the droplets could have settled. The smaller the droplets in an aerosol, the longer this takes. However, after several days, this process should certainly be complete.

The question still remains, however, how the virus makes the water leave the body. Since the away team on the Exeter is not constantly stepping in puddles, the water seems to have actually transitioned into the gas phase. Whether it evaporated (i.e., dispersed molecule by molecule into the air) or distributed itself in the form of small droplets as an aerosol in the air (as speculated earlier), we do not know. How the virus makes the host do this is another interesting question. What is truly mysterious, however, is how it manages to do this completely. As described above, a virus does not have its own metabolism. Therefore, it cannot actually do anything on its own. It always has to make its host do it for it.

Let's first take a look at the matter on the cellular level. When the infected cell has produced enough viruses, the virus could cause the cell membrane to rupture. This releases the newly produced viruses, which can then infect other cells. In the process, the leaking cell loses all its water. Well, at least if it is—literally—high and dry. Most cells (and this includes the cells in the human body) are in an environment that consists mainly of water. Destroying the cell membrane could be a significant contribution to extracting all the water from the body. However, it cannot explain the actual dehydration. Moreover, rupturing the cell membrane would not even be strictly necessary. As we saw with the salt vampire, cell membranes are semipermeable. Water can easily diffuse through them. A hole would only potentially speed up the process a bit.

The much bigger problem, however, would be osmosis. The virus from the episode *"The Omega Glory"* does, in a certain way, the exact opposite of what the salt vampire from *"The Man Trap"* does. The salt vampire extracted all the salt from the body. The water remained. The virus, on the other hand, extracts all the water from the body. Everything else, including the salt, remains. The two may do the exact opposite of what the other does. However, the problem is ultimately the same for both. It may initially succeed in extracting the substance to be removed from the body, whether it is salt or water. In the process, its concentration inside the cell decreases. The virus may, in a certain way, have an easier time. After all, water simply passes through cell membranes. Salt, on the other hand, does not. However, the transport would come to a halt relatively quickly. Otherwise, there would be a diffusion from areas of low concentration to areas of high concentration. Here, the virus has the same problem as the salt vampire.

For the virus, there is also the fact that it does not have its own body with which it could take any measures for extraction. It has to use the infected host cells for this purpose. Initially, the cells may still participate in any forced measures to release water. However, this does not go on indefinitely. This has nothing to do with the cells eventually losing interest or realizing that it is not good for them. Eventually, they simply die. This would happen well before all the water is completely extracted from the body. The cell would simply die at some point. And certainly before the last bit of water was removed. Since the virus itself cannot do anything, it is therefore not capable of actually extracting all the water from the body.

So much for the more biologically influenced aspects of this case. There is also a purely chemical question: the question of the material balance. Both the mass balance and the atomic balance pose puzzles.

First of all, it applies to every process that the total mass cannot change. A virus may trigger chemical reactions. In the process, substance A becomes substance B. Additionally, a chemical substance (water in our case) can be transported from place A to place B. Nevertheless, the total mass remains constant. What was a kilogram before is still a kilogram afterward.

Extended to the atomic balance, this means that not only the mass remains constant but also the amount of each individual type of atom. So if the virus removes all the water from the body, the number of hydrogen and oxygen atoms decreases (at least in the body itself; they are now just in a different place). All other atoms, however, remain where they are. This applies to the calcium in the bones, the iron in the hemoglobin, the phosphorus from the DNA, the nitrogen and sulfur from the proteins, the oxygen from the carbohydrates, and the hydrogen and carbon from... Well, actually from everything that occurs in the body. The human body consists of a very large proportion of water. But that is not all. There is also a lot of carbon. The fine crystals left behind by the humans do not really fit the picture. If you were to extract all the water from a human, you would not simply be left with such white crystals. There are several reasons for this.

For one, there is the matter of quantity. It is true that the human body consists mostly of water. However, the remaining substances make up significantly more than the meager pile of crystals that the Enterprise away team finds aboard the Exeter. For this, significantly more than just the water would have to be extracted from the human body.

There is also another problem. The human body is a wild mixture of thousands of different chemical substances. If you only extract the water, you do not simply get nice, white crystals. From the metals in the body (calcium, iron, sodium & Co.), such crystals could certainly be produced. Additionally, some of the non-metals could be used. Some of the oxygen, phosphate, and chloride ions could form salts with the metals. The crystals found might arise from these. But what about the carbon?

White crystals formed from carbon compounds are not entirely unusual. Just think of sugar crystals. Part of the carbon in the body is indeed bound in the form of sugars. As mentioned, crystals could form from these. However, part of the carbon is bound in substances like fats. Different people may have varying amounts of fat in their bodies. Fundamentally, however, fats make up a significant portion of every human body. We all know sugar crystals. But when was the last time you heard of fat crystals? Here, we need to look at the relationship between chemical structure and crystal formation.

A crystal is a highly ordered structure. This means that all atoms, ions, or molecules in the crystal not only have their own fixed place (around which they oscillate slightly depending on the temperature). What is also crucial: The fixed places are regularly arranged. In a salt crystal, for example, sodium and chloride ions alternate. Their arrangement is so uniform that it is referred to as a lattice. Sodium and chloride ions can be very easily assembled into a crystal. Both ions are almost spherical. Spheres can be well arranged into an ion lattice. However, if the particles become asymmetrical, it becomes more difficult to arrange them into a regular lattice.

If you look at the molecular structure of chemical substances, you can often already derive statements about the melting point. For this, you need to know two rules. First, the larger the molecule, the higher the melting point usually is. Second, the more asymmetrical the molecule, the lower the melting point. The second rule can be explained by crystal formation. Melting is nothing other than the opposite of crystallizing. The easier a substance crystallizes, the less willing it is to leave this crystal form again. The melting point is correspondingly high. Conversely, a substance whose molecules cannot be easily arranged into a regular lattice has a low melting point. The more asymmetrical molecules are, the more difficult it is to pack them into crystals. Therefore, the melting points of asymmetrical molecules are usually lower than those of equally sized, symmetrical molecules.

If molecules are very asymmetrical, it can even happen that the formation of a crystal is practically impossible. At low temperatures, such substances still seem to be solid. However, no crystal has formed. The atoms or molecules are still irregularly arranged like in a liquid. They can no longer move freely. It gives the impression that they are solid. However, a crystal is not formed when such substances solidify. This state is called amorphous. According to the strict thermodynamic definition, they are still liquid.[7]

Regardless of whether you want to consider such substances as solids or liquids, they do not form beautiful, regular crystals like those found on the USS Exeter. This is exactly where the problem with fats lies. Fats consist of a glycerol molecule to which three fatty acid molecules are bound. These elongated fatty acids are not straight rods. In particular, unsaturated fatty acids have kinks. Moreover, the three fatty acid residues from the glycerol do not necessarily all point in the same direction. A fat molecule is therefore anything but symmetrical. Additionally, there are different fatty acids. As a result, the fatty acid residues protruding from the glycerol are of different lengths. In a biological fat, the fatty acids are also randomly distributed among the individual fat molecules. Thus, you are dealing with asymmetrical molecules that are also of different sizes. This is illustrated in Fig. 4.2. How is a crystal supposed to form from this?[8] Many components of the human body may remain as crystals after the removal of water. This would not be so easily possible with fats.

Even if it were possible to crystallize all the remaining components of the body, the question still remains why this should happen. Why are all substances crystallized? And why do the individual crystals draw together into a small heap? If you remove all the water from a body, you would rather expect a kind of dried-out mummy as a result.

---

[7] This is the reason why glass is sometimes referred to as a liquid. The atoms in it are also not regularly arranged. Since glass decidedly does not form a crystal, it is strictly speaking not a solid. Since it is usually not practical to pretend that glass is liquid, there is a second definition for solids. This is not quite as scientifically clean. In this definition, a boundary in viscosity is simply defined. If the viscosity exceeds this boundary, the substance is referred to as a solid according to this definition.

[8] For this reason, the melting points of mixtures are often significantly lower than those of the corresponding pure substances. The different molecules interfere with each other during crystallization. Such mixtures, which have a lower melting point than the pure substances, are called eutectic. A well-known example of this is road salt in winter. Water melts at 0 °C. The salt melts only at several hundred degrees Celsius. However, their mixture melts even at temperatures well below 0 °C.

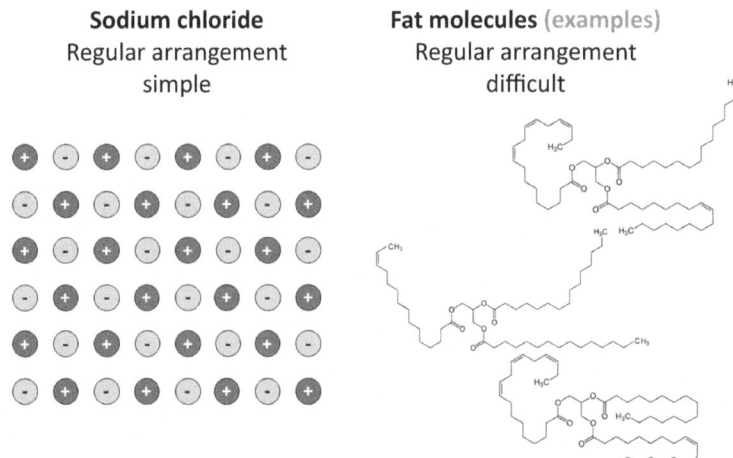

Fig. 4.2 Simplified representation of the ions or molecules in solid salt or fat; in the case of sodium chloride, the arrangement of the ions in the form of a regular lattice is quite simple, whereas crystallization is difficult with the asymmetrical and irregular fat molecules

---

Excursus

Very Fast Drying

A similar process can be observed in the 21st episode of the 2nd TOS season, *"By Any Other Name"*. In it, an away team from the Enterprise encounters a group of Kelvans. They first capture the away team and eventually gain control of the entire spaceship. To do this, they use a technical device built into their belts. When they press a button on their belt, humans are transformed into a plaster-like structure in the form of a cuboctahedron.[9] Again, it seems that all the water has been extracted from a human body.

Essentially, the same questions arise as with the crew of the Exeter. The only difference is that it happens very quickly. Within seconds, humans are transformed into handy cuboctahedrons. How long it takes in the case of the virus from *"The Deadly Years"* is unknown. However, it can be assumed that it takes much longer. Again, the question arises as to where the water disappears. Additionally, the question now arises as to how it can happen so quickly.

Two problems would arise if one wanted to completely extract the water in a fraction of a second. One of them would be more of a mechanical nature. On its way out of the body, the water would have to possess an enormous speed. In doing so, it would inevitably drag a lot along with it. In fact, it would simply tear the body apart in an explosion.

---

[9] A cuboctahedron is a polyhedron with a total of 14 faces. Of these, six are squares and 8 are regular triangles. The cuboctahedron belongs to the so-called Archimedean solids.

The second problem is at the molecular level. During the transformation, the water does not simply flow out as a liquid. Therefore, it must leave the body through the air. To do this, it must be vaporized. We will simply accept that the Kelvans' device can deliver the required amount of energy so quickly. After all, they belong to a highly developed civilization. How they apply this energy specifically to the victim is another matter. One thing is clear, though. Through this energy supply, water might be vaporized. However, it would require a rise in temperature. To vaporize all the water in a human body in a fraction of a second, one would have to supply heat at a very high temperature. Otherwise, the heat transfer into the body would not be fast enough. But even if the temperature in the body did not rise above 100°C, it would be enough to denature all proteins. So, it may be conceivable—somehow—that the Kelvans transform humans with their device into something that looks like a plaster cuboctahedron. The process would just be very difficult to realize. The reversion, however, would be truly challenging. If all proteins are denatured, then the reverted humans would certainly be dead.

An effect known from terrestrial biology can be seen in the 5th episode of the 1st DSC season, *"Choose Your Pain"*. In it, the tardigrade contracts into a small, spherical structure and loses 99% of its water in the process. Terrestrial tardigrades actually exhibit this behavior, known as cryptobiosis. They almost completely shut down their metabolism to survive in inhospitable environments. This allows them to survive extremely low temperatures or radioactivity, for example. This was demonstrated, among other things, by a space flight in 2007. Tardigrades, also known as water bears, are, however, a maximum of 1.5 cm in size. Most species are even significantly smaller. With a specimen several meters in size, the reverse problems arise that we have already encountered with 1 cm tall miniature humans. The bodies of animals with a maximum diameter of 1 cm, for example, are structured in such a way that a large part of the substance transport simply occurs through diffusion. Active pumping, like in the human heart, is not really important for such small animals. Due to the short distances, it is sufficient for substances to simply diffuse from one end of the animal to the other. If the animal grows to several meters, it suddenly has a problem. The transport of ingested food to the interior no longer works as usual. While it works for elephants, their bodies are designed to transport heat and chemical substances over several meters internally.

If giant tardigrades with a diameter of several meters wanted to switch to cryptobiosis, then substance transport would become a problem. In principle, it might be possible. The challenge, however, is the speed. Terrestrial water bears can usually switch to this state quickly enough to survive the onset of extreme conditions. A giant like the one we find on Discovery, however, would take a very long time. If the transport of water from the interior were to occur within seconds, it would probably tear the poor animal apart. ◄

4.3 Simply Being Someone Else

It would sometimes be quite practical if one could simply take on the appearance of another person. If one could simply adapt the shape of one's own body to the circumstances at any time, that would be useful. Shape-shifters are far ahead of us in this regard. Apart from a lot of mischief one could cause and some criminal options that would arise, the possibilities would be immense. What would be a real problem from the perspective of crime fighting would offer incredible opportunities for data protection.

Over time, the heroes of the various Star Trek series encounter different, mostly intelligent life forms that can take on the appearance of other living beings. As early as 1966, Gene Roddenberry had the Enterprise crew under Captain Kirk meet the M-113 creature, which would later become famous as the Salt Vampire. The crew of the Enterprise-D under Captain Picard later made the acquaintance of Allasomorphs in the 10th episode of the 2nd TNG season, *"The Dauphin."* Even the entire Voyager crew of Captain Janeway is eventually copied by the so-called Silver Blood in the 24th episode of the 4th VOY season, "Demon." Captain Archer was misled by a life form known as a Phantom, which appeared to him as a scantily clad woman, in the 18th episode of the 1st ENT season, "Rogue Planet." Our heroes also encounter shape-shifters in the Star Trek films, such as the Chameloids from *"Star Trek VI: The Undiscovered Country."* So far, there have been no real new shape-shifters in *Star Trek: Discovery,* but the integration of the Klingon Voq into the body of Ash Tyler[10], which no one can subsequently prove, also goes a bit in that direction. However, the most well-known and probably most important example of a shape-shifting species is undoubtedly the Founders from *Star Trek: Deep Space Nine.*

Odo[11], the security chief of the Deep Space Nine space station who hails from the Founders, struggles somewhat with the authentic replication of humanoid faces. However, his "more practiced" fellow beings from the so-called Great Link are already so good that they can not only appear convincingly as humans. They can also take on the appearance of any human. They are so good at it that they can deceive not only the human eye. Even the sensors that Starfleet has in the 24th century cannot detect the deception. When Odo transforms into a rock, the scanner only detects a rock. This is an impressive ability. Nevertheless, it raises a number of chemical questions again.

At the beginning of the series, Odo already demonstrates his shape-shifting abilities when he eavesdrops on a conversation between the Bajoran terrorist

---

[10] His first appearance occurs in the 5th episode of the 1st DSC season, *"Choose Your Pain."*

[11] Odo makes his first appearance in the 1st episode of the 1st DS9 season, *"Emissary,"* and subsequently in almost every other DS9 episode.

Tahna Los and the two Klingon women Lursa and B'Etor in the 3rd episode of the 1st DS9 season, *"Past Prologue,"* by transforming into a rat. This is a very practical trick, but one question remains unanswered: What happens to the rest of his mass? Very large rats may weigh half a kilogram. Most rats weigh significantly less. We do not want to speculate about the weight of Odo (and thus his now-deceased actor René Auberjonois). But it is definitely certain: He weighs more than half a kilogram.

One of the immutable laws of nature is the conservation of mass. No matter what chemical reaction takes place, no matter how much substance diffuses from one place to another, the mass remains conserved. It may be in a different place, and the chemical form may have changed. But its value remains the same all the time. The density may change. When a liquid evaporates, the vapor has a much lower density. This means its mass per volume is significantly smaller. This example makes it clear why it is so important to speak of mass and not weight in this context. Water vapor rises, while liquid water flows downward. Seemingly, water vapor even has a negative weight, which is why it rises and can even lift weights in a balloon. However, during evaporation, neither the mass nor the weight becomes negative. If you evaporate one kilogram of liquid water, you get one kilogram of water vapor afterward. Nothing changes in the mass. Even the weight does not change. The term weight refers to the force with which a certain mass of a substance is attracted by gravity. For a mass of one kilogram, this weight force is about ten newtons on Earth. During evaporation, nothing changes in the mass and thus nothing in the weight. What changes, however, is the volume. From about one liter of water, you get (depending on the exact conditions) over one cubic meter of water vapor. This displaces significantly more air, which in turn has its own weight. The weight of the displaced air is the so-called buoyancy. If this buoyant force is greater than the weight force of the vapor, it rises. This is the case with water vapor. Most other vapors (e.g., ethanol) are heavier than air under the same conditions.

Another reason why it only makes sense to speak of mass conservation and not weight conservation is gravity. The stronger the gravitational field, the lower the weight. And this, although nothing has changed in the respective body. In weightlessness the weight has a value of zero. Nevertheless, mass conservation applies not only on Earth but throughout the universe. All chemical substances must adhere to this. Regardless of whether they belong to a spaceship, a potted plant, a human, or an alien. Apparently, this fundamental principle of chemistry does not seem to apply to the shape-shifters known as the Founders. When Odo shrinks to the size of a rat, he violates a fundamental law of nature. A circumstance that is not without a certain irony, considering how much the policeman Odo values the observance of laws.

The conservation of mass is, however, only the simplest form of a conservation law that complicates shape-shifting. If one really looks at it chemically, then one must also consider the atomic balance. One must bear in mind that every body consists of specific chemical substances. This applies regardless of whether it is living or non-living matter. If a shape-shifter assumes a new form, chemical reactions could indeed occur in its body. In the process, chemical transformations can easily take place. Thus, a shape-shifter can certainly change the chemical composition of its body. However, it is strictly bound to the atomic balance. This means that, for example, it can convert a glucose molecule ($C_6H_{12}O_6$) into two alanine molecules ($C_3H_7NO_2$) to form proteins. However, it must consider what to do with the excess oxygen atoms, as two alanine molecules contain a total of only four oxygen atoms and not six like the glucose molecule. Conversely, it must find an additional two hydrogen atoms from somewhere, as the glucose molecule had only twelve, but it needs fourteen. Not to mention the nitrogen, which is not present in glucose at all, but is very much present in amino acids like alanine. If another substance is available that provides the missing hydrogen and nitrogen and absorbs the excess oxygen, then this is chemically possible. If such a substance is not available, then it becomes difficult.

The atomic balance ultimately tells us that the number of atoms of each element must be conserved. After the transformation, there must be just as many oxygen atoms as before. No more and no less. The same applies to the atoms of hydrogen, nitrogen, carbon, sulfur, phosphorus, and so on. So, if Odo transforms into a stone, then it seems that not only is there the restriction by the conservation of mass that it must be a stone weighing about 75 kg. Additionally, the conservation of the atoms of each individual element must also apply. Many stones contain a lot of silicon, which is present in very small amounts in the human body. On the other hand, biological organisms contain a lot of hydrogen and carbon, which are generally rare in stones. Every time a shape-shifter wants to assume not just a new form but a truly new shape, it will face this problem. Not only will the conservation of mass (already known from physics) set tight limits for the shape-shifter. Beyond that, chemistry with the atomic balance sets another limit regarding what shapes it can truly assume. Otherwise, the shape-shifter can only take on the outer form but would have a completely different composition inside than the person whose identity it assumes.

Let's assume that the Starfleet scanners are significantly more advanced than today's measuring devices. Then the question arises as to why they cannot distinguish shape-shifters from real humans. Otherwise, the scanners are capable of revealing all sorts of information about the internal structure of the scanned objects—including the chemical structure. So, it really seems that the shape-shifters adopt the complete chemical structure of those they copy. Which in turn raises the question of how they can transform back after the complete transformation.

If a shape-shifter transforms so completely that the 24th-century scanners, despite the greatest efforts of Starfleet, are not able to distinguish a Founder in human form from a human, then a complete adoption of the form down to the last chemical detail must have taken place. There must be nothing left inside the body that could reveal the shape-shifter as such. This consequently means that the organ controlling the transformation must also be transformed. On the one hand, the corresponding organ must change itself, which is already difficult enough. On the other hand, and this is probably much more dramatic, it is no longer present as such afterward. Therefore, it becomes difficult for the shape-shifter to transform back.

The transformation organ is overall an exciting matter. Because how does it actually effect the transformation? First of all, it must ensure that all the necessary atoms are transported to the right place in the body. Where bones are to form, there must be enough calcium. Where an eye is to form, on the other hand, there should not be too much calcium. A shape-shifter must therefore be able to transport chemical substances very quickly within its body. What challenges arise when it wants to transform as quickly as we are used to from Star Trek, we will see in the next section.

In addition to transporting substances, it must primarily control the chemical reactions. These reactions, in turn, should proceed very quickly and selectively. For this, the shapeshifter needs appropriate catalysts. The concept of a catalyst is well known. However, catalysts are not just the tube systems with platinum particles found in the exhaust system of gasoline engines to chemically convert carbon monoxide and nitrogen oxides and thus render them harmless. Catalysts are something very widespread in chemistry. Generally, catalysts are solid or liquid substances that accelerate certain chemical reactions. Fundamentally, they are not consumed in the process. If you look at the chemical processes on a molecular level, they do indeed participate in the reaction itself. The addition of a catalyst changes the reaction mechanism. This means that the intermediate steps of the reaction are different. It takes a different chemical path. This is also referred to as a new reaction pathway. If this reaction pathway is more favorable,[12] then the reaction proceeds faster with a catalyst than without. Catalysts are not only necessary for a shapeshifter to transform quickly but also to transform into the correct result. After all, only the desired reactions should occur, and as few as possible of the "wrong" reactions, which are also chemically possible.

If it is possible to specifically accelerate the desired reactions and not the undesired ones, then a reaction system can be steered in a desired direction. In technical chemistry, catalysts are very often used for this purpose. Biochemistry does the same, using (mostly) water-soluble catalysts.[13] The catalysts of biology are called enzymes. If a living being wants to change its chemical structure, it must produce

---

[12] Roughly simplified, one can imagine it as the path becoming less rocky. A certain amount of energy must be expended for a reaction to proceed. This energy is subsequently released again, but initially, it must be expended. This energy is called activation energy. If the activation energy is lower on a reaction pathway with a catalyst, an alternative path with fewer obstacles opens up. Therefore, the reaction proceeds faster with a catalyst.

[13] In biochemistry, one usually deals with so-called homogeneous catalysts. These are catalysts that are dissolved in the reaction mixture. Homogeneous catalysts can make many reactions possible even at low temperatures. They are therefore very popular in organic chemistry. Technically, they are usually very impractical because the expensive catalyst must subsequently be separated from the reaction mixture. Therefore, heterogeneous catalysts are often used in technical chemistry. These are solid bodies (like the three-way catalyst in a car) that come into contact with the reaction mixture but do not dissolve in it. Here, separation is not a problem. For biochemistry, on the other hand, this is quite impractical, as an organism would have to carry corresponding particles when using heterogeneous catalysts.

(or transport) one or more enzymes in the corresponding part of the organism. This is not unusual. In biochemistry, this happens constantly. However, unlike a complete transformation of the entire organism, only individual reactions always occur in the organism. Moreover, in most cases, these are stationary processes. This means that, for example, during respiration, sugar is continuously converted with oxygen to carbon dioxide and water, which is coupled with a reaction in which ADP (adenosine diphosphate) is converted with phosphate to ATP (adenosine triphosphate). ATP, in turn, provides the energy for all kinds of chemical processes and is converted back to ADP and phosphate. The cycle is closed, and the formation and consumption of ATP do not lead to an accumulation of ATP. It is said that the system is in a stationary state.[14] A shapeshifter, on the other hand, does not need a stationary process for transformation. It needs a reaction that purposefully and quickly converts all the chemical substances in its body into those chemical substances that should occur in the assumed form.

For this, the shapeshifter will need many different enzymes to carry out all these reactions. These enzymes must be produced very quickly in large quantities. And it must know very precisely which enzymes to use. This means that the shapeshifter can initially only transform into things it knows at least in principle. If it wants to transform into something new and unknown (to it), it first needs time to learn. It will have to find out how an enzyme must be structured to precisely accelerate the reaction needed to produce the product required for its transformation. Therefore, shapeshifters will have difficulty assuming a completely new form. For a humanoid, biological creature in the broadest sense, like the salt vampire of M-113, it may well be conceivable to take on the form of a human. However, the use of enzymes would have to be considerable to accomplish this at the shown speed. The ability of the Founders to immediately assume any shape that has nothing to do with their previous form is, on the other hand, at least enzymatically demanding.

Copying another being exactly is also difficult for a completely different reason. After all, one must first know what to transform into. Actually, that doesn't sound particularly complicated. In all scenes where a shapeshifter decides to take on the form of another living being, he has already seen the corresponding living being beforehand. So he knows what he needs to transform into. Or does he?

If a shapeshifter only wants to take on the rough outer form, then that might be enough. In some scenes, he has only seen the person whose identity he wants

---

[14] Stationarity should not be confused with equilibrium. In a stationary process, as much is produced or supplied as is consumed or removed in the same period. Therefore, the concentrations remain the same. However, there can very well be a net reaction. In equilibrium, the net reaction comes to a halt. On a molecular level, there may still be a lot of reaction. For every product molecule formed, a reactant molecule is formed again. Thus, the composition remains constant even without the addition or removal of reactants and products. For this reason, chemists never strive for their "inner equilibrium" because they know that the state of equilibrium means death (strictly speaking, even only after complete decomposition, because a corpse, although no longer in a stationary state, only slowly reaches equilibrium through decomposition).

to steal from one side. Accordingly, he only has incomplete information. We will set aside this small inaccuracy for now. Let's assume that shapeshifters have very good eyes. Then they might indeed have enough information to copy a person's appearance sufficiently to deceive another person. But the Starfleet scanners should not be fooled by this. To adapt to the template to the extent that even the ship's sensors are deceived, the shapeshifter needs much more information.

Basically, the optical impression can provide a lot of information. This includes chemistry. Light transports a lot of information about the substances from which it was emitted or reflected. Depending on which chemical substances interact with light, there can be a change in wavelength. The Raman effect is an example of this.[15] The so-called Raman spectroscopy is based on this and provides important information about substances and mixtures of substances. Moreover, it is not said that shapeshifters only see the light visible to us. If they can see slightly longer-wavelength light, then they additionally have the information of infrared light (IR). In chemistry, this wavelength range is used to elucidate chemical structures using IR spectroscopy. However, all these methods only provide information about the surface. They do not reveal what it looks like inside.

To see inside, radiation is needed that is not completely absorbed at the surface or the top millimeter of the skin. X-rays or gamma rays could help with that. However, there is darkness regarding these types of radiation in most places. This is also good because we would all get cancer if we were permanently exposed to significant X-ray or gamma radiation. Therefore, the shapeshifter would have to emit X-rays himself. It is to be assumed that the spaceship sensors would notice this. And there is another problem with X-rays and other radiations that are not completely absorbed at the body surface. The initially helpful circumstance that they pass through the body becomes a challenge here. Because they penetrate the body, they hardly interact with it. They contain correspondingly little information. You can still read from an X-ray image how the bones are distributed in the body. With very good X-ray technology, you can also recognize a bit more than that. But you don't get much information about the chemistry.

One method to obtain chemical information would be nuclear magnetic resonance spectroscopy, usually abbreviated as NMR. The NMR method provides very precise information about chemical structures and is therefore very widespread in chemistry. And even better: It can be modified to provide not only information about the chemical composition but also about the spatial distribution of substances in the body. This modification of the NMR technique is known

---

[15] In short, the Raman effect is based on an inelastic scattering of light by molecules. When atoms or molecules absorb light, they become excited. This means they transition to a state of higher energy. When they return to their original state, they emit the light. In Rayleigh scattering, they return to their initial energetic state. As a result, they emit the same amount of energy as light, and the light has the same wavelength as before. However, it is also possible for them to return to a slightly different state. Then the amount of energy released is slightly different from the previously absorbed amount, causing the wavelength to change. The wavelength change is characteristic of a substance and can thus be used for its spectroscopic identification.

as MRI (magnetic resonance imaging). However, the shapeshifter would have to rotate around the person he wants to copy (at least according to today's technology). It would be conceivable to obtain important information for copying a person in this way. However, the shapeshifter must possess sensory organs that far exceed anything biologically known to us. And by the way, he would have to generate a magnetic field. In principle, this would not necessarily have to be particularly strong at first. He could—at least hypothetically—even simply use the Earth's magnetic field—at least as long as there is such a field at the corresponding location in space. However, in chemical analysis, much stronger magnetic fields are usually used for good reason. The higher the field strength, the better the resolution. This does not only mean spatial resolution (in terms of small pixels) but above all precisely evaluable information about which chemical substances are present. If a shapeshifter wants to create an exact copy, he needs very precise information. However, the ship's sensors would probably quickly raise the alarm if a shapeshifter were to generate a very strong magnetic field on board.

A final challenge that all shapeshifters we meet in Star Trek have to deal with will be discussed in the next section. It is not enough to quickly transform substances through chemical reactions. The corresponding substances must also be transported to the right place. Achieving this quickly is not easy at all.

4.4 Why Does a Dead Shapeshifter Revert to Its Natural Form?

When I first saw the already mentioned 5th episode of the 1st TOS season, *"The Man Trap,"* there was a brief moment when I wondered. At the end of the episode, the salt vampire dies. As a shapeshifter, it had taken on the form of Nancy Crater shortly before its death. To save Captain Kirk's life, Dr. McCoy had to kill the salt vampire, appearing as Nancy, with his phaser. It then falls to the ground and remains dead. What did I wonder about? Why does it retain Nancy's form? Shouldn't it revert to its natural form now that it is dead?

Maybe one or two seconds passed, then the reversion began, and another second later, it was complete. I had wondered: Because maybe two seconds after the shapeshifter's death, it still had the foreign form. This ridiculously short duration was enough to trigger a contradiction to an expectation. A shapeshifter must revert to its original form after its death! And immediately!

I think it is a good example of how what one has seen in countless movies and series shapes one's own imagination. We assume that a shapeshifter reverts to its "natural" form immediately after its death. We have seen this in numerous episodes of Star Trek and other science fiction films. Anything else surprises us. Even if it only takes a few seconds for it to actually happen.

Why people unconsciously let themselves be so strongly influenced by what they see would certainly be an exciting psychological question. However, this book is supposed to be about chemistry. Chemistry may have little to contribute to the formation of expectations. But it has quite a bit to say about the question of whether a shapeshifter reverts after its death. If this is introduced so prominently here, then there will probably be some difficulties.

First of all, the question arises as to what actually drives the reversion. If the shapeshifter dies, then it can no longer actively revert to its original form. After all, the organ that controls and initiates the shapeshifting also dies with it. If a person dies while wearing a mask, then they still have that mask on their face after death. At least until someone removes it from the corpse's face. What should cause a shapeshifter to transform again after death? It would no longer be able to do so. Can dead matter remember its original form and revert to it?

The answer can indeed be: Yes! There are substances that can "remember" their previous form. Before homeopathy enthusiasts start to cheer, let me disappoint you right away: I'm sorry, there is no molecular memory. If you dilute a substance so much that it is no longer there, then it is no longer there and can have no effect. There is no memory of water molecules for other molecules that were once there. This (and all of homeopathy) is unfortunately just a fairy tale (with which an entire industry makes good money). The so-called shape-memory effect has nothing to do with it.

The shape-memory effect is known in both organic and inorganic chemistry. In organic chemistry, they are called shape-memory polymers, and in inorganic chemistry, they are called shape-memory alloys. How does it work?

First, let's imagine a relatively simple case of a shape-memory polymer. A solid consists of two components. One of them is elastic, the other thermoplastic. Most people are probably familiar with elasticity. A well-known example is rubber. An elastic object can be deformed. However, it constantly exerts a force to return to its original form. If the external deforming force is removed, it returns to its original form. An example of a thermoplastic material is wax. At low temperatures, it is more or less solid. At high temperatures, it becomes liquid. If you cool it down again, it becomes solid again.

Now let's imagine a material that is actually a kind of mixture of an elastic and a thermoplastic material. At low temperatures, the material is solid. If the temperature rises, the thermoplastic melts. Due to its close mixture with the elastic material, the melted thermoplastic cannot flow away and remains distributed inside. Since the thermoplastic material is now liquid, the elastic material can be deformed by an externally applied force. If you cool the whole thing down again, the thermoplastic becomes solid. If you have maintained the deforming force until then, the body could not be returned to its old form by the elastic component. Once the thermoplastic component has solidified, the external force can be removed, and the body remains in its new form. The elastic component wants to return it to its original form, but the solidified thermoplastic prevents this. However, if the temperature rises again, the thermoplastic becomes soft or even liquid again. The elastic component can now return the body to its original form.

This form of the shape-memory effect is called the one-way memory effect. In the two-way memory effect, there are two forms that the material permanently remembers. One for high and one for low temperatures. In some materials, the

4.4 Why Does a Dead Shapeshifter Revert to Its Natural Form?

transformation to the original form can also be triggered by effects other than temperature. In some substances, the effect can be triggered by magnetic fields or light of certain wavelengths. How exactly the reversion after death is supposed to be triggered in a shapeshifter remains a mystery that still needs to be explored in the infinite expanses of space. However, it is fundamentally possible for a shapeshifter to revert after death. Whether it is a necessity that its body consists of a shape-memory material is another matter. However, it is not entirely far-fetched. If the shapeshifter wants to completely take on the copied form, then its shapeshifting organ must also transform. To return to its natural form, a shape-memory effect would at least be an interesting option. However, the requirements for such a shape-memory material exceed everything we know in this area today.

The second aspect to consider in the reversion after death is the speed. This is not only a problem if the shapeshifter reverts to its original form within seconds after its death. It is also a problem if it wants to take on a new form within seconds while alive. This requires substances to move very quickly from one part of the body to another and to get to exactly the right position there.

We know very rapid transformations of this kind from Star Trek not only from shape-shifters. An important example comes from the medicine of the future. The so-called dermal regenerator ultimately does nothing different than a shape-shifter, which takes on a new form within seconds.[16] The dermal regenerator is one of the medical devices that almost everyone would sometimes wish for. If a patient's skin is severely or slightly injured, Starfleet doctors resort to exactly this device. It is aimed at the injured area from about ten centimeters away. At the push of a button, a beam emerges that makes the wound disappear within seconds. The device is even so good that it simply makes blood drops sticking to the skin disappear. Within a few seconds, the injured Starfleet officer is back in perfect health. The wound is gone, and everything is fine. Who wouldn't wish for that today?

Why is it unfortunately not quite so simple in practice? If you get injured nowadays, the best you can get is a cream that accelerates healing. But it is not about healing within seconds. A wound cream may shorten a multi-day (or multi-week) healing process by a few days. This has little in common with the medical miracle of the dermal regenerator. Why has no one invented such a device to this day? And why is it unlikely that anyone will invent something like this anytime soon? And what does this have to do with dying shape-shifters?

In both cases, we experience a very rapid substance transport. This means that chemical substances are transported from one part of the body to another. Why should this be a problem? After all, humans can move not only a few centimeters within seconds but up to several meters in a second.

When a person moves their arm or leg, they can indeed do so very quickly. However, this movement process is quite simple for one reason: Everything moves in the same direction. The skin moves from place A to place B. The muscles move

---

[16] Dermal regenerators appear in numerous episodes of Star Trek; one example is the 11th episode of the 4th DS9 season, *"Homefront"*.

from place A to place B. The bones move from place A to place B. Simply everything that belongs to the arm or leg moves from place A to p lace B. There is no serious resistance to overcome. The path is practically clear. The only thing in the way is some air. You can feel this during fast movements. Especially when driving fast, air resistance becomes quite noticeable. Even when a person moves their arm very quickly, the air resistance is ultimately manageable. Some air has to flow from place B to place A. That's basically all there is to it. The reason why the resistance to movement through the air is so low lies in a property called viscosity. Viscosity is something that all liquids or gases possess. Simply put, it is the thickness. Honey has a fairly high viscosity, and water has a rather low one. Accordingly, the resistance is much greater when you try to stir honey than when you stir water. The concept of viscosity can be applied to gases like air in the same way. Their viscosity is very low, and correspondingly, the resistance of the air to movements within it is low.

What does this have to do with the transformation of shape-shifters and dermal regenerators? Well, when a shape-shifter takes on a new form (this includes returning to the "natural shape"), a lot of movement takes place. This often goes unnoticed at first. When a shape-shifter changes from the shape of one person to another, the external shape remains largely the same. Similarly, the dermal regenerator does not seriously deform the patient from the outside. However, in both cases, considerable movements must occur inside the body. And the difficulty is that the path is not clear.

Let's consider the process using the example of the dermal regenerator. We are dealing with a piece of skin that has been injured by some external impact. The dermal regenerator must now cause several movements simultaneously. First of all, all contaminants that occur in a fresh wound must be removed. In other words: They must be moved out. Here we already have the first difficulty because the question arises as to where. Since the contaminants must be removed from the body, but the dermal regenerator apparently does not suck them up and absorb them, only the air remains. All contaminants must therefore be vaporized. To vaporize something, you normally heat the thing to be vaporized. Therefore, you would have to deal with very high temperatures at least locally. This might even be helpful for disinfecting the wound. At the same time, however, the proteins of the still intact body cells would denature. This is referred to as the coagulation of proteins. The manufacturers of dermal regenerators must ensure that their devices do not make the injury worse instead of healing it.

The transport of cells and the extracellular matrix then becomes really complicated. Whether a dermal regenerator repairs skin or a shape-shifter changes its form, cells always have to be moved to the right place. And not only that. The same must happen with the extracellular matrix. What is this about? The human body is not just an accumulation of cells. If it were, it would simply fall apart, to put it simply. Because what would hold the cells together? Therefore, the biochemistry of humans includes the entire area between the cells. Here, the extracellular matrix holds the cells together. The extracellular matrix not only serves the function of holding the cells together but also acts as a water reservoir, among other things.

4.4 Why Does a Dead Shapeshifter Revert to Its Natural Form?

Chemically speaking, the extracellular matrix is a collection of macromolecules that do not directly belong to a cell. Macromolecules are molecules that, unlike oxygen or water, do not consist of two or three atoms but rather hundreds or even thousands. Often, macromolecules are chain-like in structure. They do not simply form a lump of interconnected atoms. Instead, atoms are arranged in one (or often several) rows. A well-known example is our genetic material. This is stored in the form of deoxyribonucleic acid (DNA). The DNA forms the well-known double helix. Many atoms are arranged in two rows. Each of the two chains is not just a row of atoms but has small branches. The atoms of these branches form, among other things, a connection to the second chain. As a result, the two chains are not free but closely bound together. Because the two chains form parallel spirals (helices), it is called a double helix.

DNA is just one of many examples of macromolecules. Another example is starch. In it, many sugar molecules are lined up in a long chain. The most important type of macromolecules in our context are proteins. These macromolecules, built from chains of amino acids, perform a multitude of functions in the body. One of these is the construction of the cell membrane, the envelope of the cells. The function important to us in this context is holding the cells together. Here, a group of proteins called collagen plays a significant role. As connective tissue, collagen essentially holds the cells together. And this is exactly where our problem lies. We not only have to transport cells, but we also have to transport them through collagen (aside from the fact that collagen probably also needs to be transported). Collagen was historically used to make glue. This makes sense because collagen's function in the body is precisely to hold things together. However, it also highlights the problem: we need to move cells through a mass that is chemically designed to hold cells in place. This might work. But one thing it certainly is not: fast.

When regenerating skin or shape-shifting, not only entire cells need to be transported. Sometimes, it is just individual substances. One could imagine, for example, that amino acids need to be transported from point A to point B in the body to form proteins for skin regeneration or shape-shifting. Due to their chemical structure, amino acids are quite water-soluble and can thus be easily transported. However, this also takes time. After all, we do not just want to move a few molecules a few micrometers or perhaps millimeters. This is where we transition between two disciplines of chemistry: organic chemistry on one side and technical chemistry on the other.

Organic (as well as inorganic) chemists are capable of synthesizing all sorts of things chemically. There is hardly anything they cannot produce. However, there is one problem. Organic chemists like to produce a few milligrams of a substance. Sometimes they produce a few grams. If you ask your colleagues in organic chemistry for a few hundred grams, it becomes difficult, not to mention tons. The methods with which organic chemists carry out their—truly remarkable—chemical syntheses work quite well on the scale of milligrams or perhaps a few grams. If you want to produce larger quantities, entirely new problems suddenly arise. Therefore, there is another discipline of chemistry called technical chemistry.

Technical chemistry deals with producing in large quantities what organic, inorganic, or biochemistry creates in the lab.[17] One of the major challenges in this is the transport processes. If you are producing only a few milligrams, it is not a problem. However, if larger quantities need to be processed quickly, entirely new effects arise that were not noticed in the lab. One transport process is heat transport. For example, the reaction heat must be removed from the reaction site. If you work in the lab with very high dilution of the reaction mixture in a solvent, it is not a problem because the reaction medium hardly heats up due to the reaction heat. If you work at higher concentrations, the mixture heats up significantly more. Additionally, if you have a larger quantity, the paths the heat must take are much longer. This leads to local overheating. Technical chemistry takes care of this. Similarly, technical chemistry deals with mass transport. The reactants must reach the reaction site. In our case: amino acids must reach the place where proteins are needed to restore damaged skin. We are not dealing with a package of amino acids that can simply be moved from A to B. Even in this case, we would have the aforementioned problem of having to move through the rather tough collagen. However, amino acids are not transported in a package but individually. And in chemistry, individually means truly individually. Each molecule must reach its destination individually.

The transport of molecules that are not packed in a larger package occurs through diffusion. Anyone who wants to observe the speed of substance transport through diffusion should drop a piece of sugar into their coffee and wait. If you do not stir, the coffee is usually cold before the sugar has evenly distributed in the cup. This is precisely the problem of the dermal regenerator. If the mechanism of diffusion has to bring the substances needed for the regeneration of injured skin to the right place, it takes time. The transport of the necessary substances to the right place is one of the reasons why a wound does not heal within seconds but takes days or weeks. The speed of substance transport is one of the main difficulties in building a dermal regenerator. And it is the great challenge for any shape-shifter who wants to take on a different form quickly and cannot take days or weeks.

To make the whole story even more complicated, substance transport presents another problem. What should actually cause the molecules to move not only quickly but especially in the right direction? How does a dermal regenerator or a shape-shifter get substances to diffuse to where they are supposed to go?

Let's think back to the sugar in the coffee cup. The sugar dissolves over time. Chemically speaking, its molecules leave the solid structure of the sugar crystals, dissolve in the water of the coffee, and diffuse away from the sugar crystal. In the long run, the sugar distributes itself evenly in the cup. What we cannot observe,

---

[17] There is a specific field of study for this, called chemical engineering. Chemical engineers are responsible, among other things, for technical chemistry, i.e., producing large quantities with reasonable effort. Additionally, chemical engineers ensure that the final product is pure. Chemical syntheses often occur in solvents that need to be separated. Moreover, there are often by-products or unreacted starting materials that need to be separated. Chemical engineering deals with all these issues.

however, is that the sugar molecules suddenly simply wander back to the sugar crystal. The cause is the concentration gradient mentioned earlier. The molecules diffuse from an area of high concentration to an area of low concentration. Not the other way around. The direction of diffusion is not controllable. It simply follows the change in concentration. Always from the area of high concentration to the area of low concentration.

If a dermal regenerator is to quickly repair defective cell membranes and other structures through the synthesis of appropriate proteins, it needs very large amounts of amino acids directly at the site of protein synthesis. The amino acids would therefore have to accumulate in certain places. In other words: places with a high concentration of amino acids would have to be created. However, from a place with a high concentration, the respective substance diffuses away. The last thing that would happen is that the amino acids diffuse to a place in the wound where their concentration is higher than in the rest of the body. If they do not do this, then the wound cannot be healed within seconds. Here, chemistry makes life difficult for both the designer of a dermal regenerator and a shape-shifter who wants to transform quickly.

New Materials in the 23rd and 24th Century

5.1 How many Elements are there Actually?

Verterium, Tritonium, Veridium, Element number 247. The list of chemical elements encountered in Star Trek is long. Now, the list of elements encountered in daily life is also not exactly short. We breathe oxygen and additionally draw nitrogen into our lungs and out again. We often handle things made of iron or aluminum. We know metallic objects coated with zinc or chrome. We often hear about hydrogen and carbon. We use portable electronic devices in which batteries based on lithium are installed. Ultimately, all matter is made up of atoms, and each atom can be assigned to an element. Often a substance consists of atoms of several elements. When different atoms form a common molecule, we call it a compound. Different metal atoms that form a metallic material are called an alloy. Many substances we deal with daily are not pure substances, even if they look like it. Such mixtures that look like pure substances are called homogeneous. But even mixtures can be traced back to elements. These are partly mixtures consisting of different elementary substances (like air, for example). Partly, they are mixtures consisting of compounds (like saltwater, for example). But even that can all be traced back to elements.

How many elements are there actually? And is there an upper limit to their number? Can there be an unlimited number of (previously unknown) elements?

Classical chemistry was known to assume four elements: earth, water, air, and fire. The idea of four elements was not only prevalent in the Middle Ages but extended well into modern times. However, it was developed in antiquity. This concept already had precursors in ancient Greece.[1] The so-called pre-Socratic

---

[1] The three philosophers mentioned below came from Miletus and Ephesus. These places are not in Greece but are all in present-day Turkey. Culturally, however, they belonged to the Greek cultural sphere at that time.

philosophers assumed a single primordial substance from which everything derives. Thales considered water to be this primordial substance, and Anaximenes considered air to be it. Heraclitus, on the other hand, assumed that fire must be the primordial substance. His reasoning was that everything in the universe is constantly changing. Fire seemed to him to best correspond to this, which is why he declared it the primordial substance.

In the fifth century BC, Greek philosophers like Empedocles finally developed the doctrine of the four elements from this. Among others, Plato and Aristotle contributed to its further development. The idea behind the four-element doctrine was that different substances have different properties. Some are solid. Others are liquid. Some are gaseous. For each of these states, it was thought, an element was needed. Solids can melt, as is well known. The idea was that such substances consist of earth and water. Today, we no longer describe solid, liquid, and gaseous as manifestations of individual chemical elements but as states of matter. Incidentally, fire also fits into this scheme. The so-called plasma is sometimes referred to as the fourth state of matter. A plasma is initially a gas. However, the individual atoms of this gas are (usually due to high temperature) broken down into electrons and atomic nuclei or atomic cores (an atomic core is an atom that has only given up part of its electrons). Flames have a plasma-like character. In this respect, the four classical elements actually represent the states of matter quite well.

The idea of only four elements has a certain charm. It provided a model that seemingly could explain the entire chemistry. Everything newly discovered could somehow be traced back to the four elements. But even in antiquity, there were considerations to expand the circle of elements. Thus, not only the alchemists searched for the fifth element, the quintessence ("quinta essentia", the fifth being). All these thoughts are extremely fascinating from a history of science perspective. However, they have little to do with the actual functioning of chemistry.

As we know today, matter consists of atoms. The first thoughts in this direction also go back to Greek philosophers of antiquity. The modern concept of atoms, of which there are different kinds, was then developed, starting in the 17th century with Robert Boyle, by scholars like Bernoulli and Dalton. But how many kinds of atoms are there actually?

First of all, the number of types of atoms is very large. However, two different types of atoms do not necessarily belong to two different elements. An atomic nucleus (or nuclide) is initially determined by two parameters. The number of protons and the number of neutrons. Generally, as the number of protons increases, the number of neutrons also increases. This is not a necessity, of course. There are indeed atomic nuclei that have the same number of neutrons but differ in their number of protons. Such nuclides are called isotones. Conversely, there are atomic nuclei that have the same number of protons but differ in the number of neutrons. Such nuclides are called isotopes. Additionally, there is the case where two atomic nuclei differ in both the number of protons and the number of neutrons. However,

5.1 How many Elements are there Actually?

the sum of both (the so-called nucleon number) is the same. Such nuclides are called isobars. In one of these three cases, the atoms belong to the same element. In which one?

The considerations regarding protons and neutrons belong to nuclear chemistry. Often, the question of the structure of atomic nuclei is simply assigned entirely to physics. This is somewhat simplified, although chemistry generally does not concern itself much with the structure of atomic nuclei. What interests chemistry are the bonds that exist between atoms. Accordingly, the chemical bonding properties determine which atoms belong to the same chemical element and which do not. Which of these two factors influences these bonds between atoms? The number of neutrons or the number of protons?

The chemical bond is mediated neither by protons nor by neutrons. The cause of chemical bonds is that electrons distribute themselves in specific ways between atomic nuclei. If there is a concentration of electrons between atomic nuclei, then one is dealing with a so-called covalent bond. If electrons do not distribute themselves in a small area between two atomic nuclei but rather across the entire solid body, this is referred to as a metallic bond. If there is a redistribution of electrons such that one of the atoms has "too few" and the other "too many" electrons, the oppositely electrically charged atoms attract each other. This is called an ionic bond. In any case, it is always the electrons that determine the chemical properties. Therefore, atoms with the same number of electrons should be counted as the same element. The neutrons in the atomic nucleus hardly influence the electrons at all. Thus, the number of neutrons is also irrelevant. On the other hand, protons influence the electrons significantly. The positively charged protons attract the negatively charged electrons. To be electrically neutral, an atom must have exactly one electron per proton. Therefore, the number of protons determines the number of electrons and thus the chemical properties. Two atoms with the same number of protons have the same chemical properties. Therefore, they belong to the same chemical element.

The number of chemical elements is therefore significantly smaller than the total number of types of atomic nuclei. Almost all elements have different isotopes. However, the number of neutrons cannot be varied arbitrarily. If the atomic nucleus contains too many or too few neutrons per proton, it becomes unstable. This means that it is transformed into another nuclide through radioactive decay. The ratio of neutrons to protons increases slightly with increasing proton number. Thus, in heavy elements, there is on average more than one neutron per proton. However, this does not change the fact that the number of isotopes of an element is limited. In some elements (such as fluorine), only a single isotope occurs naturally. The number of elements, on the other hand, is not fundamentally limited.

The number of protons can initially be any natural number. Starting from 1 (hydrogen) to 2 (helium), 3 (lithium), and so on, until one eventually reaches 92 (uranium). Then it stops for now. No more elements are found in nature. All heavier elements can only be artificially produced.[2] The artificial production of these elements, called transuranium elements, is not only quite complex. There is another problem: they are unstable. They are radioactive and decay into lighter elements. Normally, they do this by emitting alpha radiation. Although there are radioactive isotopes of almost every element, up to and including element 82 (lead), there is almost always at least one stable isotope. All elements that contain more protons than lead are radioactive with all their isotopes.[3] The higher one goes in the atomic number, the more unstable the elements become. This is why no transuranium elements are found on Earth. Perhaps there were heavier elements than uranium on Earth during the formation of the solar system. However, these have long since decayed radioactively. Therefore, transuranium elements must be artificially produced. In the case of the next heavier elements, this can still be done in a conventional nuclear reactor. If uranium is bombarded with neutrons, not only can nuclear fission occur, but heavier elements can also be formed. For example, elements 93 (neptunium) and 94 (plutonium) are produced in nuclear reactors. If the correct plutonium isotopes are hit by neutrons, they first transform into the next heavier plutonium isotope and then through beta decay into an isotope of element 95 (americium). If plutonium nuclei are hit by alpha particles, element 96 (curium) can be formed. One might already suspect that the process becomes increasingly difficult.

At some point, the production of elements as a byproduct in a nuclear reactor comes to an end. For even heavier elements, one must eventually resort to particle accelerators. Simply put, these are (very large) devices that shoot individual nuclides at each other so that they collide at very high speeds. High speed means a lot of energy. This way, new, even heavier elements can be created. It becomes apparent that the process becomes increasingly complex. Additionally, one does not obtain many atoms of the new element when the atomic nuclei have to be shot at each other individually.

In addition to the significant effort required for production, the new elements are becoming increasingly unstable. Even if they can be produced, it does not take long before they decay again. While the most stable uranium isotope (element 92) still has a half-life of several billion years, for element 95 (americium) it is only a few hundred years. For element 100 (fermium), it is barely a hundred days. For element 105 (dubnium), even the most stable isotope has a half-life of only about

---

[2] Strictly speaking, this is not entirely correct. The elements with numbers 93 (neptunium) and 94 (plutonium) do indeed occur naturally on Earth in tiny traces. However, the quantities are so small that these two elements are generally considered artificial elements.

[3] In older literature, bismuth, element number 83, is also still considered a stable element. This can be explained by the fact that the most stable bismuth isotope is indeed radioactive, but its half-life is significantly longer than the current age of the universe. Therefore, the statement that there is a stable and thus non-radioactive bismuth isotope is formally incorrect but practically correct.

half a minute. For element 110 (darmstadtium[4]) the most stable isotope seems to have an average lifespan of a whole minute. For element 115 (moscovium), one should expect the decay of the atomic nucleus within less than a second.

In principle, the number of elements is not limited. Mathematically, the number of protons can be increased arbitrarily. The elements just become increasingly unstable. The new elements that appear in Star Trek are usually used as materials to make special devices. However, this becomes somewhat difficult if half of the element has already decayed radioactively within seconds or even fractions of a second. Is there perhaps still a chance to find more stable new elements?

From the shell model of the atomic nucleus (analogous to the shell model of the atom, which plays a significant role in chemistry), it can be predicted that certain combinations of proton and neutron numbers result in particularly stable atomic nuclei. In the case of particularly stable transuranium elements, this is referred to as "islands of stability." The corresponding nuclides should be particularly stable. For the element with atomic number 114, such an island of stability is predicted. And indeed: flerovium, the name of the element, has proven to be particularly stable. Its most stable isotope has a half-life of at least 5 seconds. Compared to its direct neighbors in the periodic table, which at best have a lifespan of a few hundred milliseconds, this is quite a lot. This flerovium isotope has 171 neutrons in addition to its 114 protons. For the flerovium isotope with 184 neutrons, which has not yet been produced, significantly higher stabilities are predicted. However, it seems very unlikely that a technically usable material can be obtained from a flerovium isotope. Even if the half-life were a few days or even years, it would still be far too little for practical use as a material.

But let's just assume that there are more elements that are truly stable. After all, there is no real upper limit to the number of protons in an element. If it is possible to find more stable elements, then one could certainly do great things with them. What exactly can be done with them can only be said to a limited extent now, as we do not yet know the chemical properties of these new elements. Incidentally, we do not even know the chemical properties of most of the transuranium elements we already synthesised very well. This is ultimately not really surprising, because how can you conduct chemical reactions with something of which you have only produced a few individual atoms? And these atoms then exist for only a few minutes, seconds, or even just milliseconds. Certain predictions about their chemical properties can be derived from their position in the periodic table of elements. Elements that stand below each other there have similar properties. Thus, it can be assumed for element 118 (oganesson) that it is probably not very reactive, as it belongs to the eighth main group. The elements above it are all noble gases.

---

[4]Yes. The element with atomic number 110 is indeed called darmstadtium and is named after the Hessian city of Darmstadt. The German *Helmholtz Center for Heavy Ion Research* (formerly *Gesellschaft für Schwerionenforschung*) is located there, where a number of other elements were first produced in addition to darmstadtium. From this location, element 108 also received the name hassium (Latinized for *Hesse*).

Unfortunately, more than such speculations about the chemical properties are not available.

If new, stable elements are possible, then they can also be produced. An advanced civilization would have to build enormous particle accelerators for this purpose. The real challenge would probably not be the fundamental production. The real difficulty would be to realize such a process in a way that allows significant quantities to be produced. This is probably the reason why we never hear about facilities in Star Trek where new elements are produced. Instead, there are always mining facilities where such elements are extracted from foreign celestial bodies. Or—which also happens frequently—the scanners discover a new element somewhere. In both cases, the new elements naturally occur. Can additional chemical elements naturally occur somewhere in space?

To answer this question, we first need to look at how naturally occurring elements actually come into being. Let's start at the very bottom. With the element that has a single proton and no neutron in the atomic nucleus: hydrogen. Well, hydrogen is the only element where the atomic nucleus does not have to form as such. It is simply a proton. Naturally, the proton also had to come into being somehow. However, this is a completely different process by its nature. The question leads back to the creation of the universe and is quite exciting in itself. The formation of elementary particles such as protons, however, has nothing to do with chemistry. Chemically speaking, we only need to ask how protons become hydrogen. This can be explained by the fact that protons are positively charged and electrons are negatively charged. Therefore, they attract each other. A single proton sooner or later captures a single electron. Together they form a hydrogen atom. When the hydrogen atom encounters another hydrogen atom, they form a chemical bond. A hydrogen molecule is formed. This is hydrogen as we know it. This is how the path from a single atomic nucleus to a "real" chemical element works for all elements. The remaining question is then only how the atomic nuclei of the other elements come into being.

In the case of helium, this is quite well known. The nuclear fusion inside the sun constantly converts large amounts of hydrogen into helium. Therefore, helium is the second most common element in the universe after hydrogen.[5] But why does nuclear fusion not continue in many stars? Why do helium nuclei in the sun not continue to fuse with each other (or with hydrogen nuclei)? Why does the process stop at helium and not continue to lithium, beryllium, and so on?

This is because helium is very stable. This applies not only to its (non-existent) chemical reactivity. It also applies to its nuclear chemical properties. Helium does not undergo radioactive decay and it almost as reluctantly fuses with other atomic nuclei to form heavier nuclides.

---

[5] The helium with which we fill balloons on Earth, however, does not come from the sun but from natural gas. During radioactive decay inside the Earth (specifically: alpha decay), a helium nucleus is emitted by the decaying atomic nucleus. The helium nucleus captures two electrons and accumulates (provided it does not escape into the atmosphere and eventually into space) as helium in natural gas.

5.1 How many Elements are there Actually?

The formation of helium through nuclear fusion in the sun works so well because it is energetically at a significantly lower energy level than hydrogen. Therefore, a lot of energy is released during the fusion of hydrogen to helium. This initially applies to the fusion of almost all light elements. However, helium is an exception. If helium is further fused to lithium or beryllium, no energy is released. If one were to further fuse lithium or beryllium nuclei, energy would be released again. The process initially stops at helium, which has a very low energy level. Energy would have to be put into its fusion. Only carbon (element number 6) then has a lower energy level than helium. However, one can only reach this through several fusion steps from helium. Therefore, nuclear fusion hardly progresses beyond helium. Stars primarily produce helium.

However, this does not mean that it cannot continue at all. At a certain point in their development, stars begin to fuse helium into heavier elements. Once the path to carbon has been taken, energy is ultimately released again during fusion. Further elements can be formed. At least up to a certain point. At a proton number of 26, we reach the most stable element: iron. This metal is not only mechanically very stable. It is also extremely stable in nuclear chemical terms. If one were to fuse iron atoms into heavier atoms, energy would have to be supplied again. Therefore, elements with proton numbers less than 26 tend to undergo nuclear fusion. Elements with proton numbers greater than 26 tend to undergo nuclear fission. Energetically speaking, atomic nuclei try to approximate iron. Therefore, elements with larger atomic numbers do not actually form through nuclear fusion. Well, at least not in the first place! Because somewhere the 66 elements must come from, which occur in nature and have more protons in the atomic nucleus than iron.

If heavier elements are bombarded with neutrons, they can transform into heavier elements through subsequent beta decays. In a beta decay, a neutron in the atomic nucleus transforms into a proton. It emits an electron (called a beta particle) and an antineutrino. This way, heavier elements can slowly form. Even beyond the 26-limit of iron. This happens to a small extent in stars. However, really large amounts of heavy elements do not usually form this way. And very much heavier elements than iron also do not result from this process. Really large amounts (and really heavy elements) can form in supernovae. The unknown, stable elements that the starfleet's spaceships repeatedly encounter in space could have formed in a super-supernova. No, that is not a typo. It is indeed meant to say "super" twice. What one is to imagine under such a double supernova, I do not know myself. But it is clear that it would need something significantly more intense than a supernova as we know it to form new stable elements.

The fact that no stable transuranic elements have been found in nature to date shows that a simple supernova cannot form them. It may well be capable of forming transuranic elements. However, since their half-lives are significantly shorter than the age of the solar system, they have already decayed to the point where their remnants are below the detection limit. If the known processes for forming heavy elements really produced stable transuranic elements, they would still be present. Since they are not, it obviously requires something significantly stronger than a supernova. Whatever that may be: It has not occurred in our part of the galaxy so far.

> **Excursus**
>
> **Biological Origin of Elements**
>
> In the year 2371, the USS Voyager arrives at a planet with a ring system. In this ring system, they find a new element. The proton number of this element, at 247, is already impressive enough. Even more impressive is how this element is formed. The German title of the 9th episode of the 1st VOY season does not refer to it, but it fits quite well: *"Das Unvorstellbare"* (i.e. *"The Unimaginable"* or *"Unthinkable"*).
>
> As it turns out, this element is formed during the decomposition of the corpses of the Vhnori. Their dead are placed on asteroids that form the planet's ring system. This is a truly remarkable process. Decomposition is a process in which small organisms like bacteria chemically transform a dead body. In a chemical reaction, the elements themselves do not change. If the body consisted of X atoms of carbon, Y atoms of oxygen, Z atoms of hydrogen (and so on) at the time of death, the decomposition products still consist of X atoms of carbon, Y atoms of oxygen, Z atoms of hydrogen (and so on). The elements are found in completely different chemical compounds after decomposition. Tendentially, smaller molecules are formed during decomposition than were originally present in the living organism. The elements may also be distributed differently in space (some, for example, transition into the gas phase). However, the overall composition, at least in terms of the elements, remains unchanged.
>
> The question indeed arises as to how the decomposition of the Vhnori is supposed to form a new element. Do bacteria with built-in particle accelerators live on the asteroid? That would certainly be a fascinating biological discovery. It seems more likely, however, that it is not a biological decomposition at all. Theoretically, it is conceivable that there is radiation on the aforementioned asteroids that triggers an element transformation. We would then not be dealing with biological decomposition, but with (previously unknown) radiation-induced chemical decomposition.
>
> Three questions still remain: First, which (probably very heavy) element is present in the body of the Vhnori and can serve as the starting point for the formation of element 247? Second, if such radiation exists on the asteroid, why did the Voyager's scanners not detect it? And third, why did the away team survive this radiation? The intensity of the (e.g. neutron) radiation should be lethal. ◄

5.2 Materials That Do Not Consist of Chemical Elements

All substances are composed of elements. Sometimes it is only one element. Most of the time, it is several. The individual atoms are held together by chemical bonds. Intermolecular interactions exist between the individual molecules, which,

5.2 Materials That Do Not Consist of Chemical Elements

for example, prevent a liquid from evaporating or plastic from melting. All these bonds are formed through electrical attraction. Specifically, it is the attractions between the positively charged atomic nuclei and the negatively charged electrons. The electrons distribute themselves in such a way that a bond is formed. This applies both to the bonds between atoms in a molecule or metallic solids and to the interactions between different molecules.

These chemical bonds can be very strong. Just think of a diamond. Through the covalent bond between carbon atoms, it becomes the hardest known material. But could there be even more stable materials?

Even the strongest chemical bond between atoms has only a certain strength. With diamonds, the potential for strength through chemical bonds is largely exhausted. Admittedly, this is already quite a lot. Chemical bonds can indeed be very strong. Nevertheless, a maximum achievable strength is reached, simply put, when the strength of all chemical bonds in a cross-section is added up.[6] Since atoms are very small, there are very many bonds between atoms in a cross-section. This is how the remarkable strengths of materials like diamond or carbon fibers come about. However, the known chemical bonds cannot offer much more. The strength of materials, regardless of which elements they consist of, cannot become infinitely large.

But there must be more. Is there not some material whose strength is even higher than what today's known chemical bonds allow? The heroes from Star Trek occasionally encounter a substance so stable that even the most modern weapons of the 23rd and 24th centuries cannot damage it: Neutronium.

Even Captain Kirk encountered it in the 6th episode of the 2nd TOS season, *"The Doomsday Machine"*. The hull of the unmanned weapon system known as the Doomsday Machine is made of Neutronium. This makes it completely immune to any attack. Since it is a planet killer (and the system also carries out its purpose quite indiscriminately), this proves to be a serious problem. Finally, the Doomsday Machine can only be stopped by flying a spaceship the size of the Enterprise into its opening. Fortunately, the USS Constellation is available for this suicide mission.

Over a hundred years later, Kira and Garak have an encounter with Neutronium in the final episode of the series, the 26th episode of the 7th DS9 season, *"What You Leave Behind, Part II"*. To end the war with the Dominion, they need to break into the central command on Cardassia Prime. Since they do not have a key, they want to try using a bomb to open the door. However, it turns out that the door in question is made of Neutronium, and a bomb would not even cause a dent. Therefore, they need another plan.

---

[6] This is indeed very simplified. One must consider that there are different criteria for strength. It can make a significant difference whether one pulls on a material, compresses it, or twists it. The idea of the sum of all bonds over a cross-section would only be an approximation of tensile strength. It does not say much (at least directly) about other strength parameters.

In the course of Star Trek's history, there are a whole series of encounters with neutronium or neutronium alloys. But what exactly is neutronium? A look at the periodic table of elements reveals no element by that name.

Chemical elements are characterized by the interplay of protons in the atomic nucleus and electrons in the atomic shell. This is what determines the chemical properties. The neutrons are rather incidental. Let us imagine an atomic nucleus whose number of protons is zero. If the number of neutrons is now greater than zero (otherwise there would be nothing at all), then one would have an atomic nucleus that contained no protons. The corresponding atom would therefore also have no electrons. Since neutrons have no electric charge, they cannot bind electrons. This has two consequences: On the one hand, chemical bonds as we know them are impossible. Without electrons to mediate the attraction between positively charged atomic nuclei, none of the known types of chemical bonds work.[7] On the other hand, the distances between the atomic nuclei could become very small. Normally, atomic nuclei do not come arbitrarily close. This is simply because the positively charged protons in the atomic nuclei repel each other. If the atomic nucleus consisted only of a neutron, then this repulsion would not exist. The "atomic nuclei" could aggregate into an incredibly dense material. This material is called neutronium.

The idea of such a substance and the naming as neutronium is almost a century old. The German chemist Andreas Antropoff, who was born in Estonia, took the periodic table (developed by Meyer and Mendeleev) in the 1920s and did something similar to what we did in the last section. In the search for new possible elements, we mentally extended the periodic table upwards (in terms of atomic number; as they are usually printed it is downwards). In doing so, we considered what happens when the atomic number (the number of protons) increases further and further. Antropoff considered what happens when the atomic number is reduced below 1 (hydrogen). In 1925, he submitted an article to the journal *Angewandte Chemie*, which was published the following year under the title "Eine neue Form des periodischen Systems der Elemente" ("A New Form of the Periodic System of Elements"). In it, he proposed a modified arrangement of the elements in the periodic table. This did not catch on. However, it had a peculiarity: there was still a space left to the left of hydrogen. This circumstance led him to speculate about a hypothetical element with atomic number 0. For the atom, he proposed the term neutron. For the corresponding element, he suggested the name neutronium.

To this day, it has not been possible to prove such an element. It is now generally accepted that neutrons exist. Both as a component of atomic nuclei and as

---

[7] In the 2nd episode of the 1st season of PIC, *"Maps and Legends,"* an android responds to the question of what is brown and sticky with, among other things: "Boson-enriched nanopolymer." How to enrich a substance, whether a nanopolymer or something else, with bosons is still completely unclear by today's standards. However, bosons are those elementary particles that mediate forces between fermions (which include electrons, protons, and neutrons). A completely new type of bond could therefore be present in boson-enriched nanopolymers. How to imagine this is left open. But it would at least explain why it is so sticky.

5.2 Materials That Do Not Consist of Chemical Elements

individual elementary particles, their existence is undisputed. In atomic nuclei, we actually have the situation that several neutrons—as in hypothetical neutronium—are connected. However, protons are always present as well. Much has been speculated about neutrons that connect without a proton being involved. Such a thing has not been proven to this day. Theoretical calculations also suggest that it would be unstable. The strength of the bond between the individual neutrons would probably even be negative. But what kind of bond should hold neutronium together at all? After all, there are no electrostatic attractions between differently charged particles in it.

In nature, there are fundamentally four types of interactions. Interaction can mean both that things attract each other and that they repel each other. It can also involve a combination of different forces. In such cases, attractive forces predominate at a greater distance and repulsive forces at a smaller distance. This is the principle of a bond. Bonds are designed to hold particles together (this is the attractive force that predominates as long as the particles do not come too close). At the same time, bonds ensure that the particles do not completely collapse into each other (these are the repulsive forces that predominate when the particles come too close). In chemistry, forces are always mediated by electrical interaction. From everyday life, we know a second interaction: gravity. This holds solar systems together as gravity and prevents us from simply jumping into space. In addition, there are two other fundamental interactions. The strong and the weak interaction. The strong interaction holds atomic nuclei together. The electrical interaction between the positively charged protons would otherwise immediately tear the atomic nucleus apart. The weak interaction, in turn, is responsible for atomic nuclei experiencing radioactive decay. The strength of all these forces decreases with increasing distance. Electrical interaction and gravity decrease with the square of the distance. This means that at twice the distance, only a quarter of the force acts. This initially sounds as if these two interactions would not be effective over very large distances. As we can see from gravity, which holds solar systems and even galaxies together, this is not necessarily the case. In the case of the strong and weak interaction, the force decreases even more significantly with increasing distance. This means that the strong interaction usually hardly extends beyond the atomic nucleus. For chemistry, it is therefore usually not particularly interesting. Except in the case of an "element" like neutronium.

In neutronium, the distances between the "atoms" are not, as usual, on the order of a few picometers. They are only a few femtometers.[8] One can imagine a piece of neutronium in a certain way as a single gigantic atomic nucleus. The strong

---

[8] While even the smallest atom, hydrogen, has a diameter of a "gigantic" 50 picometers (a pico is a 1 with eleven zeros before the decimal point), its nucleus, consisting of a single proton, is only about 1.5 femtometers in size (a femto is a 1 with 14 zeros before the decimal point). The hydrogen atom is thus more than 30,000 times larger than its nucleus—if one expresses the difference in distances. In terms of volume, it is even more than 30 trillion times larger than the proton inside it.

interaction could theoretically hold neutronium together accordingly. A material held together by this short-range but unimaginably strong force should be fabulously stable, at least in terms of its mechanical strength.[9]

The mere fact that many neutrons are brought together does not necessarily mean that the strong interaction really holds them together. Let's think again about the heavy elements from the last section. The more nucleons present in an atomic nucleus, the more unstable it becomes. It is therefore at least more than questionable whether a material like neutronium can exist and whether it really has the fabulous mechanical properties we see in Star Trek. However, there is another effect that could make neutronium very strong—even if the strong interaction does not: gravity.

Neutronium would have an unimaginable density. The order of magnitude of the density of this material would be about ten to a hundred trillion times what we otherwise know. "Heavy as lead" is no longer an appropriate metaphor. There is simply no comprehensible comparison for it. Due to the enormous density, you have an incredible amount of mass in a comparatively small volume. An object like a door or even a gigantic spaceship, like the planet killer, made of neutronium would have the mass of a smaller celestial body (in the case of the planet killer, the mass would even correspond to a fairly large celestial body). Now, asteroids or small planets are known not to have an enormously large gravitational pull. But that is because they are again too large. Their mass is manageable compared to large celestial bodies. However, their diameter is still several kilometers (or even hundreds of kilometers). We remember that the strength of gravity decreases with the square of the distance. If the mass of a small planet is concentrated to the size of the entrance door of the Cardassian Central Command, then the distance is very small. An enormous gravitational field would emanate from every part of the neutronium door. The strength of the attraction in gravity depends again on the mass of the attracted object. Since the other parts of the door are also made of neutronium, they would also have a huge mass. Accordingly, they would attract each other very strongly. The neutronium door would simply be held together by its own gravity. However, it could happen that it deforms into a sphere due to its own gravity. This would be the same effect that brings planets and stars into a spherical shape. The architect would therefore have to take special care of this.

Although we do not yet know whether a chemical element neutronium can really exist, and we do not really know exactly what its properties would be and what kind of bonding would cause them, we can already address two practical

---

[9] What an alloy of neutronium with another element would look like remains completely unclear. The strong interaction may connect a large number of neutrons to form neutronium. An alloy is a combination of several metallic elements. The type of bond is the so-called metallic bond, in which the electrons are distributed over the entire metallic body (hence the good electrical conductivity of metals). Alloys (like all metals) are held together by the electrical attraction between these negative electrons and the positive atomic nuclei, thus obtaining their strength. How the metallic bond is supposed to work in conjunction with the electrically neutral neutrons remains unclear.

problems. The first is again gravity. This force may hold neutronium together. But gravity does not only act between the individual neutrons. It also acts between the neutrons and their surroundings. If you build a door with the mass of a small planet, then the question arises not only of which door opener should open this megaton-heavy door. There is also the problem that everything else is attracted in the same way. The mass of a small planet at a few meters distance would create such an attraction that anyone who comes near the neutronium door would be so strongly attracted that they would simply stick to it and barely get away again.

The second problem is the production of neutronium. Neutrons can be formed from protons in two ways: by emitting a positron or by capturing an electron. An advanced civilization might be able to bring this about deliberately. Compacting the neutrons into neutronium would then be even more demanding. After all, neutrons have no charge. Neutron radiation is therefore poorly steerable. And even if they have managed to get this problem under control, there remains the question of where to get the raw material from. After all, we are talking about entire asteroids that would have to be processed even for smaller objects made of neutronium. The artificial production of neutronium would therefore hardly be practical. So, it would have to be found and mined somewhere in nature. Just as all chemically used elements are found in nature and not artificially produced. Such a natural source of neutronium could indeed exist.

At the end of the life of very massive stars, black holes are formed. They even have a higher density than neutronium (if one still wants to speak of density in the case of black holes). Smaller stars like our sun eventually become white dwarfs. Chemically speaking, white dwarfs consist mainly of carbon and oxygen. If the star has a bit more mass than our sun but is too small for a black hole, then something else can form at the end of its life: a neutron star.

After the nuclear fusion in the star has come to an end, no more energy is released. As a result, the radiation pressure, which has so far opposed gravity, decreases. The star collapses under its own weight. Consequently, everything inside is enormously compressed. In the process, the temperature rises sharply again. Finally, the protons inside begin to capture the electrons. In doing so, they transform into neutrons (and emit a neutrino).[10] A celestial body is formed that essentially consists of neutrons. Hence the name neutron star.

In a way, such a neutron star thus consists of neutronium, and one would only have to establish a mining colony there. If it weren't for two problems again. One

---

[10] The capture of electrons by neutrons was speculated about as early as the beginning of the 20th century. Antropoff also dealt with this question. In a 1924 article titled "Zur Umwandlung von Quecksilber in Gold" ("On the Transformation of Mercury into Gold"), he discusses this possibility. In the end, he concluded that this would not be possible in most cases because certain isotopes would have to occur in nature, and argon in the air would have to be transformed into these chlorine isotopes by electrons during lightning strikes. In this article, he ultimately concludes that the production of gold from mercury would be the only transformation of a chemical element in this way that could not be ruled out at that time. Today we know that even the production of gold in this way is not feasible. That would not be chemistry, but alchemy.

is (once again) gravity. Landing on a neutron star might still be possible. Working there without being crushed by one's own body weight would be virtually impossible. Subsequently, taking off against gravity with a cargo hold full of neutronium would be utterly unthinkable.

The second problem is the rotation of the neutron star. Here, the inclined reader might object that planets also rotate and one can still land on them. That is correct to some extent. However, a planet like Earth rotates only once a day on its axis. For a neutron star, it is several hundred times. Per second. The speed at the surface near the equator is several tens of thousands of kilometers per second. Therefore, a landing would only be feasible near the poles. However, since the centrifugal force, which counteracts gravity somewhat, is absent there, the problems with gravity would become significantly more pressing.

Neutronium may indeed possess a grandiose strength. However, there are so many problems that one understands why this miracle material is ultimately used very rarely even in the future of Star Trek.

Excursus

Gold-Pressed Latinum

After this discussion of materials that are not based on chemical elements, we want to return to the chemical elements once more. Especially to one of the elements that we only know from Star Trek. It is supposed to be an element that the Ferengi value above all: Latinum.

For the Ferengi, profit is everything, and latinum is their currency. In their lives, they are guided by the Rules of Acquisition. This is a collection of maxims that the Ferengi uphold with almost religious fervor. In the 16th episode of the 4th DS9 season, *"Bar Association,"* we learn that Rule of Acquisition 263 states: "Never allow doubt to tarnish your lust for latinum." And from the 20th episode of the 5th DS9 season, *"Ferengi Love Songs,"* we know: "Latinum lasts longer than lust" (Rule of Acquisition 229).

Compared to worthless gold, latinum is particularly coveted. Reason enough to take a closer look at this element. Latinum is quite obviously an element. How do we know this? I can't think of an episode where it is explicitly stated that it is an element. Ultimately, however, it can only be one. Latinum serves as currency because it is not replicable. Any chemical compound, on the other hand, would be replicable. A compound can be made from other compounds or from the underlying elements. In particular, stable compounds are usually easy to produce. Latinum is obviously very stable. Apart from the fact that it would otherwise be unsuitable as a basis for currency, we learn in the 26th episode of the 2nd DS9 season, *"The Jem'Hadar,"*: "Nature decays, but latinum lasts forever" (Rule of Acquisition 102).

This property can ultimately only be possessed by latinum if it is an element. A compound would hardly be so stable and not replicable. That it is an element can also be deduced from the name. The ending -um indicates a

5.2 Materials That Do Not Consist of Chemical Elements

metallic element. Most chemical elements are metals. This is especially true for the heavier elements. Since we do not know latinum to this day, it must be a very heavy element. Its atomic nucleus would have to contain at least 119 protons according to current knowledge[11]. This high number of protons makes the stability somewhat questionable again. On the other hand, it fits with a statement from the 16th episode of the 1st VOY season, *"Learning Curve."* In it, the Bolian Chell complains that his backpack is so heavy as if it were full of latinum bars. The fact that the atomic nucleus contains many protons (and thus also neutrons) does not necessarily mean that the density of an element is high. It first means that the weight of the individual atoms is high. However, if the substance is not a gas, one can usually assume that heavy atoms lead to high density.

A unique feature of Latinum is its state of aggregation. It is liquid. Almost all metals are solid at room temperature. There is only one exception: mercury. Like Latinum, it shines silvery and melts already at $-38.8°C$. Unfortunately, mercury is quite toxic. For this reason alone, mercury is not really suitable as a means of payment. Latinum, on the other hand, seems to be largely non-toxic. At least for Lurians. In the 12th episode of the 6th DS9 season, *"Who Mourns for Morn?"*, the Lurian Morn stores pure, liquid Latinum in one of his stomachs. This does not exactly speak for dangerous toxicity.

The fact that it is liquid, on the other hand, is quite impractical for handling. For this reason, the Ferengi press Latinum into worthless gold. Accordingly, it is mostly referred to as "gold-pressed Latinum." How this pressing is done is unfortunately never explained. It is probably not a process under high pressure, but simply refers to the fact that Latinum is bound to gold. One can indeed assume a kind of chemical compound. In principle, it would be conceivable that liquid Latinum is contained in a chamber inside a gold bar. However, a relatively soft material like gold does not seem to be really suitable for this. The risk of leakage would be too great. Moreover, gold would only cause unnecessary weight that one would have to carry around. If it were just a kind of capsule, then one would most likely have chosen another material. Even if gold might be worthless and therefore cheap to obtain in the future, other materials would simply be more sensible for this.

A compound of Latinum and gold seems much more plausible. Latinum can be integrated into the lattice structure of gold. One does not even have to really press. Melting is basically enough. If you mix liquid gold and liquid Latinum and let it solidify again, you get an alloy. As long as the Latinum content is not too high, the resulting alloy would be solid at room temperature (as long as

---

[11] The fact that it must have at least 119 protons is derived from the fact that the heaviest (at the time of writing this book) known element, oganesson, has 118 protons. Realistically, it must be even more. The (not yet produced) element 119 would be an alkali metal and thus very reactive. However, latinum appears to be rather inert in Star Trek.

there is not eutectic). It would consist of gold and Latinum atoms, all of which have a fixed position in a crystal lattice.[12]

The incorporation of one element into the crystal lattice of another element can occur in various ways. On the one hand, the foreign atoms can be located in the interstices between the host atoms. Light elements, with small atoms, sometimes embed themselves in the atomic lattices of heavier elements in this way. On the other hand, the foreign atoms can take the positions of host atoms. In this case, Latinum atoms would be found in some positions where gold atoms would otherwise be. Since Latinum is an unknown and therefore logically heavy element, the latter variant seems much more plausible. The Latinum atoms are likely simply too large for the interstices between the gold atoms.

Since the Latinum content in gold-pressed Latinum seems to be very low, the bars would largely look like gold bars. This raises the question of authenticity testing. It seems hardly conceivable that the Ferengi would rely solely on an embossed seal from a trusted issuing authority. After all, Rule of Acquisition 239, as we learn in the 25th episode of the 4th DS9 season, *"Quark's Bar"*, states: "Never be afraid to mislabel a product." The Ferengi must therefore have some way to check whether the worthless gold really contains Latinum. How they do this, we do not learn. However, it does not necessarily have to be particularly difficult.

When another element is introduced into the atomic lattice of a metal, the electronic structure changes. These changes are often very slight. However, they can sometimes be quite significant. Just think of the doping of semiconductors. This is, in a way, the chemical basis of electronics. When tiny amounts of phosphorus or aluminum are added to silicon, its electrical conductivity increases significantly.[13] A foreign atom in one hundred thousand host atoms can already have a considerable effect. Although we actually know nothing about the electronic structure of Latinum, it can be assumed that it has a significant effect on some easily testable property of gold. Otherwise, authenticity testing would be very difficult. This would be very unlikely. After all, Ferengi businessmen certainly do not let themselves be easily deceived. ◄

---

[12] The term crystal should not be confusing in this context. Crystals are not only the sparkling, angular things known from mineralogy. A crystal is generally referred to when the individual atoms have fixed positions that are arranged according to a strict order principle.

[13] However, one should not add both elements at the same time. Phosphorus causes a so-called n-doping in silicon. In this process, additional electrons introduced by the phosphorus contribute to conductivity. Aluminum leads to p-doping. In this case, the absence of an electron increases conductivity. Instead of an electron, the electron hole (called a hole) is transported during current conduction. If both elements are added to silicon, the two effects cancel each other out. The additional electron from the phosphorus fills the gap caused by the aluminum.

5.3 The Mixing Ratio of Matter and Antimatter

For the energy supply of Starfleet ships, antimatter seems to play a major role. Every elementary particle has an antiparticle. There is an antiproton for the proton, an antineutron for the neutron, an antielectron (better known as a positron) for the electron, a corresponding antineutrino for each neutrino, and so on. A special case is the photon, the "light particle." It is its own antiparticle. The universe consists practically only of matter. Antimatter is almost non-existent in the universe. This is quite fortunate because when matter and antimatter meet, they immediately annihilate each other and convert into energy; from our current perspective, a very dangerous circumstance. On the other hand, it could serve as a tremendous energy source, which would certainly be advantageous for interstellar spacecraft.

Now someone might argue that antimatter is more a question of physics and not chemistry. That is not entirely wrong. Nevertheless, it is worth looking at the topic through the eyes of chemistry. Some things that appear clear from a physical perspective become relative.

In the 19th episode of the 1st TNG season, *"Coming of Age"*, we gain insight into an entrance exam for Starfleet Academy. One of the questions posed to the applicants there deals with the correct mixing ratio between matter and antimatter in a spaceship's propulsion system. Parameters such as the type of ship, the distance of the journey, and the size of the tanks are given. The answer is ultimately: one to one. Because the amount of matter that is converted by antimatter is exactly equal to the amount of antimatter. The task with its detailed specifications was a trick question. It must always come down to this mixing ratio. It does not matter what type of ship it is, how far the journey goes, or how full the tanks are. Physics gives a clear answer here. Chemistry, on the other hand, says: It is not necessarily that simple.

To understand the problem, it is worth considering the chemical basics of combustion a bit. The combustion of hydrocarbons still forms the backbone of energy technology today. Just as antimatter reactions might in a few centuries. Let's take the combustion of natural or biogas as a simple example. Their main component is methane, and we will just assume that our fuel consists only of methane. The combustion then proceeds according to the following chemical reaction equation:

$$CH_4 + 2\,O_2 \rightarrow CO_2 + 2\,H_2O$$

Each methane molecule is converted with two oxygen molecules into one carbon dioxide molecule and two water molecules. This numerical ratio is called the stoichiometry of the reaction. The stoichiometric mixing ratio between methane and oxygen is therefore one to two. It always requires exactly twice as much oxygen as methane. In practice, fuels are rarely burned in pure oxygen but rather in air. This is cheaper and more practical. Dry air consists of about 21% oxygen. In reality, air is rarely truly dry. There is always some water vapor present. However, this water content is not constant but subject to certain fluctuations. Especially at high temperatures, it can become quite high. As a result, the proportions of the other gases in the air are lower. For the sake of simplicity, we will now just assume that the air consists of 20% oxygen. On the one hand, this accounts for the fact that there is water vapor in real air. On the other hand, it makes the calculation significantly easier (20% oxygen content corresponds to exactly one-fifth of the total air volume). If you need two liters of oxygen per liter of methane[14], then you need two times five, that is, ten liters of air. This ratio is thus the stoichiometric mixing ratio. It corresponds somewhat to our one-to-one mixing ratio of matter and antimatter.

However, technical combustion processes are almost never realized with a stoichiometric mixing ratio. Instead, a significant excess of air is usually used. This is called lean combustion. The air excess is described by the air number Lambda. This indicates the ratio of the actual amount of air to the stoichiometrically required amount of air. If you mix air and fuel exactly in the stoichiometrically correct ratio, then Lambda is equal to one. If you burn a lean mixture, i.e., with an excess of air, then Lambda is greater than one. With a rich mixture, which would correspond to a lack of air, Lambda would be less than one. But why is a lean mixture almost always used in practice? What is the reason for the excess air?

When burning a hydrocarbon, it is converted with oxygen into carbon dioxide and water. At least in the case of complete combustion. In reality, the reaction does not proceed completely. Therefore, the product is not simply a mixture of carbon dioxide and water (plus nitrogen, which is present in large quantities in the air but does not participate in the reaction). The product mixture also contains residues of the unconverted fuel and oxygen that did not oxidize these residues. Even those parts of the fuel that were oxidized are not necessarily fully oxidized. In complete oxidation, each carbon atom takes up two oxygen atoms. Carbon dioxide (CO_2) is formed. In partial oxidation, the carbon atom takes up only one oxygen atom. Carbon monoxide (CO) is formed.

Such an incomplete combustion is undesirable. On the one hand, the energy content of the fuel is not fully utilized. On the other hand, carbon monoxide and methane are significantly stronger greenhouse gases than carbon dioxide. Thirdly, carbon monoxide is toxic. Therefore, it is important to achieve as complete a combustion as possible. One way to do this is with an excess of air. If the air ratio Lambda is increased above the value of one, then more than one oxygen molecule is available for each carbon atom of the fuel. This reduces the probability that the carbon atom will not be oxidized or only partially oxidized. This achieves a more complete combustion. By using the excess air, it simultaneously happens that the oxygen is not completely converted. Basically, however, this is more or less irrelevant. The air was simply taken from the environment and is returned to it. As much as the release of carbon dioxide into the air may be a problem, the intake

---

[14] Strictly speaking, stoichiometry gives us the molar ratios, i.e., the ratio of the number of molecules of the individual substances. However, if we consider the substances as ideal gases, which is a reasonable assumption here, then it holds that at the same temperature and pressure, equal volumes contain the same number of molecules. Therefore, we can easily replace the molar ratio with the volume ratio.

5.3 The Mixing Ratio of Matter and Antimatter

and release of air is irrelevant. The only disadvantage of a high excess of air is that the additional oxygen (and with it the accompanying nitrogen) must be heated. This may slightly reduce the energy utilization. However, this is a small price to pay to avoid incomplete combustion.

Now back to the mixing ratio of matter and antimatter. To speak in terms of combustion, antimatter represents the fuel. Matter corresponds to the oxidizer, i.e., the oxygen. If a larger amount of antimatter is allowed to meet matter, then it is important to avoid the analogue of incomplete combustion. The mutual annihilation of matter and antimatter, unlike combustion, does not proceed as a chemical reaction. Chemical reactions take time. This is one of the reasons that lead to incomplete combustion. In the reaction of matter and antimatter, this limitation does not exist. However, another effect should be considered. A lot of energy is released during the annihilation. This applies to combustion and even more so to matter-antimatter annihilation. This causes the mixture to push apart. In extreme cases, it leads to an explosion. If it is possible to control the whole thing, then not everything will necessarily blow up immediately, but the released energy still contributes to increasing the distances between the particles. This makes their encounter less likely. In the case of combustion, this contributes to incomplete combustion. In the case of the matter-antimatter reaction, it means that not all antimatter particles are neutralized by matter particles. Even if it is possible to compress the beams of matter and antimatter with some force fields so that they do not expand, not every antimatter particle will necessarily collide with a matter particle. Therefore, some of the antimatter will not be neutralized.

The fact that a kind of "exhaust gas" is created, in which unreacted matter is present, is initially not very problematic. The fact that an exhaust gas with unreacted antimatter is created, on the other hand, can develop into a serious problem. Physically, it may be absolutely correct that the ratio of matter to antimatter must be one to one. Technically, however, it is not quite so simple. To solve the corresponding problems, it could be an option to orient oneself to chemistry and use an excess of matter in the antimatter energy systems, analogous to the excess air in combustion. Otherwise, there is a risk that unreacted antimatter will remain. This will subsequently annihilate with matter. However, this will not happen inside the reaction chamber, where it is supposed to happen. Instead, uncontrolled annihilations (with corresponding energy releases) outside the reactor are to be feared. Then it does not matter whether it is a Galaxy-class ship, how far the journey is supposed to go, and how big or full the tank is—the ship would probably not survive the uncontrolled energy release anyway. So, if the candidates for Starfleet Academy indicate a certain excess of matter in the exam, the corrector should actually accept it (at least as long as a good justification is provided, which admittedly would be a challenge due to the time pressure).

Particularly Impressive Chemicals

6.1 Corbomite or Kirk's Favorite Chemical Bluff

When Captain Kirk finds himself in a hopeless strategic situation, he often tries to change the game. When he no longer has a reasonable next chess move, he resorts to poker. In other words: he bluffs. He uses this trick twice and claims that the Enterprise has a chemical substance called Corbomite on board.

What is Corbomite supposed to be? The term is obviously meant to be a chemical trivial name. In addition to using trivial names, chemical substances can be named systematically. These systematic names have the advantage that one knows the chemical structure of the substance as soon as one hears or reads the name, even if one does not actually know the substance yet. Let's take the substance 2-Isopropyl-5-methylcyclohexanol (Fig. 6.1) as an example. The name means that the molecule is based on a cyclohexane molecule. These are six carbon atoms (the Greek word component "hex" means six) that form a chain. This chain is, in turn, closed into a ring (hence "cyclo"). Attached to this cyclohexane molecule is a hydroxyl group. The ending "ol" tells us this. Additionally, there are two other functional groups on the ring: an isopropyl group and a methyl group. Their exact position in the molecule is indicated by the numbers. Counting begins at the hydroxyl group. The position on the cyclohexane ring where this is attached is given the number 1. From there, one counts the individual positions on the ring clockwise. Since the isopropyl group is at the next position, it is given the number 2. There is nothing at the next two positions, so the methyl group gets the number 5. At the last position, number 6, where one is again in direct proximity to the hydroxyl group on the ring, there is also no other functional group. Therefore, this number, like 3 and 4, is not assigned.

This methodology of systematic names is very helpful. Chemists do not have to remember quite as many names, and there is immediately a name for every newly discovered substance. However, names like 2-Isopropyl-5-methylcyclohexanol

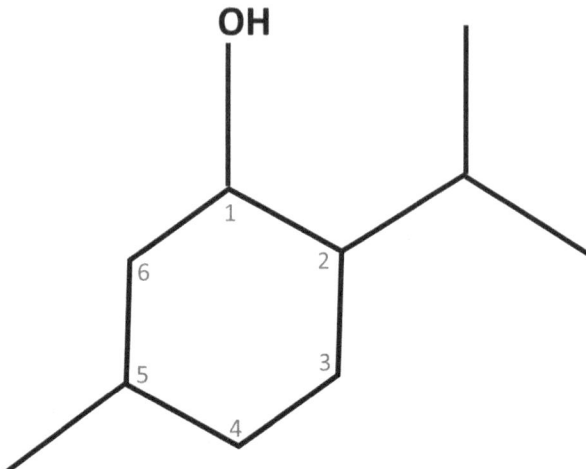

Fig. 6.1 Structural formula of 2-Isopropyl-5-methylcyclohexanol (Menthol); at the endpoints of each line, unless another type of atom is indicated, there is a carbon atom plus as many hydrogen atoms as needed for the carbon atom to be connected to exactly four atoms; the numbers in the ring are included as a guide to explain the counting principle for the position designation

are not really practical. Therefore, many commonly used substances have short names. These are called trivial names. In the case of 2-Isopropyl-5-methylcyclohexanol, the substance can be much more easily referred to by simply saying menthol. If one does not know what menthol is, one cannot deduce anything about its chemical nature from the trivial name. On the other hand, one only has to remember a much shorter name. Additionally, writing and talking about chemical substances is significantly simplified.[1]

Captain Kirk does not provide a systematic name for Corbomite but only uses the trivial name. For good reason: he does not know it himself, as Corbomite does not exist at all. He just made up the substance. The whole thing is a bluff, and he is indeed betting quite high with it. The claims he makes about the imaginary Corbomite are very bold.

Already in the first season of the original series, he uses his invention to escape from a hopeless situation. In the 3rd episode of the 1st TOS season, *"The Corbomite Maneuver"*, the Enterprise unknowingly enters the territory of the so-called First Federation. The said First Federation is not very pleased about this

---

[1] Anyone who finds the systematic name of menthol already long (or not long enough) should take a look at the systematic name of Penicillin F: (2S,5R,6R)-6-[[(E)-hex-3-enoyl]amino]-3,3-dimethyl-7-oxo-4-thia-1-azabicyclo[3.2.0]heptane-2-carboxylic acid. It becomes clear why trivial names are so important in chemistry. And this is just the tip of the iceberg. Systematic names can become arbitrarily complicated, especially in biochemistry.

6.1 Corbomite or Kirk's Favorite Chemical Bluff

and now threatens to destroy the Starfleet ship. To prepare for death, the crew is given ten minutes. As time progresses and all other options are exhausted, James Kirk resorts to bluffing. He claims that the Enterprise carries a substance called Corbomite. This Corbomite would release the same amount of energy if the Enterprise were destroyed, thus also destroying the attacking ship. Balok, the commander of the First Federation's ship, apparently takes the bait. He is at least uncertain enough to refrain from destroying the Enterprise for the time being. Eventually, Kirk manages to resolve the conflict peacefully and even establish diplomatic contacts with an exchange program. The key to this was initially his bluff with the Corbomite.

Of course, Balok did not know whether Captain Kirk was telling the truth or just trying to deceive him. But could he have known?

The question arises as to what would happen if Corbomite actually existed. If it released as much energy as was used to destroy the Enterprise, one still wonders why that should destroy the Fesarius (Balok's ship) as well. How should Corbomite reflect the energy back to the Fesarius in a targeted manner? The energy would more likely disperse evenly in all directions in the form of an explosion. And even if the energy completely hit the attacking ship, the Fesarius is many times larger than the Enterprise. The chances should not be so bad that it can withstand much more. If Balok had thought a bit more about the physics behind Kirk's claim, he might have realized that it was an empty threat. However, this book is not about physics but about chemistry. Could a bit more basic chemical understanding also have led Balok to see through the bluff?

Apparently, Balok is not familiar with the Born-Haber cycle (or whatever this fundamental natural law of chemistry might be called in his culture). This principle is normally used to determine the energy turnover of individual reaction steps. We don't need to consider it that complicated here. For us, only one essential basic assumption is important: From a chemical reaction, you get exactly as much energy out as you would put into its reverse.

Let's imagine this with a simple example illustrated in Fig. 6.2. Let's take a simple alkane like hexane. Hexane burns very well. In doing so, it reacts with oxygen and forms carbon dioxide and water. As is generally known, fire is hot. In other words: Energy is released in the form of heat during combustion. The reaction is exothermic. The amount of energy released in the reaction depends on the specific reaction. If you burn 1 mol of hexane (which corresponds to about 86 g), an amount of energy of 4.2 megajoules is released (which is a bit more than one kilowatt-hour). As products of the reaction, 6 mol of carbon dioxide and 7 mol of water are formed.

Hexane and oxygen can be obtained again from carbon dioxide and water. This is not quite simple. Chemically, you have to go through many steps. Plants do something similar. In photosynthesis, they convert carbon dioxide and water into sugar and oxygen (which again happens in many individual steps). To be able to do this, they have to supply energy to the reaction. In the case of photosynthesis, this happens in the form of light. A glucose molecule essentially differs from a hexane molecule in that it contains six oxygen atoms. In photosynthesis, not all

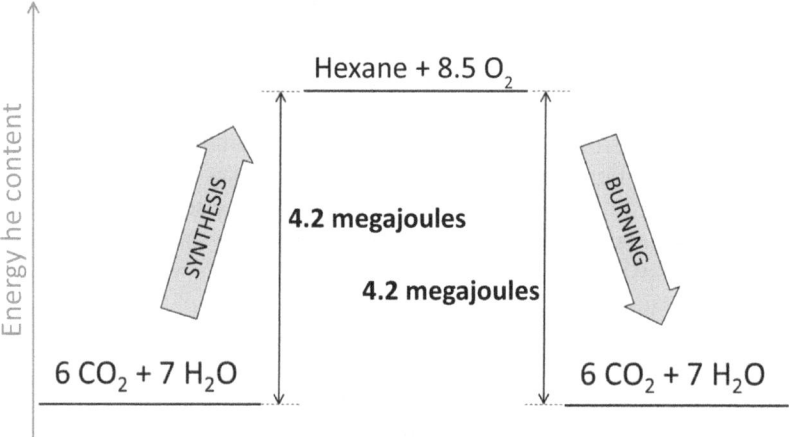

Fig. 6.2 Energy turnover in the synthesis of hexane from carbon dioxide and water and the subsequent combustion; the energy turnover of the forward and reverse reactions is the same (only the sign differs)

the oxygen is removed from the carbon dioxide. To get from glucose to hexane, you need quite a bit more energy. If you sum up the total amount of energy that you need to put in net to convert 6 mol of carbon dioxide and 7 mol of water into 1 mol of hexane and 8.5 mol of oxygen, you get 4.2 megajoules. Exactly the same amount of energy that was released during combustion.

It does not matter at all which path you take to get back to hexane. You could go through glucose and then convert it into hexane. Alternatively, you could split the water by electrolysis into hydrogen and oxygen. You could then further react the hydrogen with the carbon dioxide in a reaction known as Fischer-Tropsch to hexane. Additionally, a lot of water would be produced, which you could feed back into the electrolysis. Another variant could be that you do not split the water by electrolysis but thermally. In this process, water is decomposed by the supply of heat at very high temperatures. Ultimately, it does not matter how you do it; only one thing is crucial for us: The same amount of energy that is released in combustion must be supplied to the synthesis. This does not mean that no heat could be released in any of the synthesis steps. The Fischer-Tropsch reaction, for example, is quite exothermic. So, heat is released. The heat that has to be put into the thermal water splitting, for example, compensates for this again. The net heat supply is the same. This independence of energy turnover from the reaction path is known as Hess's law. In the end, in our hexane cycle, it is always 4.2 megajoules of heat that must be supplied net during synthesis. And it is 4.2 megajoules that are released during combustion.

A substance like the imaginary Corbomite would certainly not simply burn. At least not with elemental oxygen. That would be far too slow to destroy an

6.1 Corbomite or Kirk's Favorite Chemical Bluff

attacking ship. For explosives to react explosively, the oxygen is directly contained in the explosive. In the case of TNT (trinitrotoluene), it is the three nitro groups that provide oxygen directly for the reaction. In the case of classic gunpowder, which Captain Kirk mixes together in the fight against the Gorn in the 19th episode of the 1st TOS season, *"Arena"*, it is the saltpeter. Chemically, saltpeter is ultimately nothing more than a nitrate salt. Both the nitro group in an organic molecule and the nitrate ion in an inorganic salt are a compound of oxygen and nitrogen. This is quite unstable. If it decomposes, it provides oxygen. If this happens inside the fuel, no oxygen has to be laboriously transported from the outside. The reaction can therefore proceed abruptly and cause an explosion. It would have to be similar with Corbomite. Because if the reaction proceeded slowly, the energy released could hardly destroy the attacking ship.

What exactly Corbomite is supposed to be chemically and what reaction it is supposed to perform remains open. Presumably, even Captain Kirk has no idea. The real weak point in Kirk's bluff, however, is the matter of the energy balance. For every megajoule put into the synthesis, you get a megajoule out in the reverse reaction. That fits perfectly with Kirk's claim that Corbomite would release exactly the same amount of energy that was expended to destroy the Enterprise. Or does it?

It would work if Corbomite were formed during the enemy's attack. For this, the entire energy of the attack would have to be absorbed and Corbomite formed with it. Alternatively, Corbomite could also be decomposed by the energy of the attack. It all depends on which reaction absorbs energy and which releases energy. The formation of Corbomite or its decomposition? The energy-absorbing (endothermic) reaction must occur first. Secondly, the reaction must run backward to release the energy. Let's just assume that Corbomite decomposes when absorbing energy and reforms when releasing it. Conversely, the argument would ultimately work the same way.

First of all, there is an upper limit to the amount of energy that Corbomite could absorb. Decomposition requires a certain amount of energy. This is essentially the reaction enthalpy, i.e. the difference in energy level between Corbomite and its decomposition products. Additionally, some energy might be needed to bring the Corbomite to reaction temperature. Heating also costs energy. The more energy supplied, the more Corbomite can be decomposed. Simply put, double the amount of energy means double the amount of Corbomite. The maximum absorbable amount of energy is thus determined by the amount of Corbomite. Kirk's claim that Corbomite would absorb and release all the energy of the enemy's attack is therefore on shaky ground. It may be true, but only up to a certain point, which is determined by the amount of stored Corbomite. For the 23rd century, however, this seems very questionable. Just consider that in the future, antimatter weapons like photon torpedoes are used by starships. The amounts of energy released are enormous. Therefore, a huge amount of Corbomite would be needed. Although Kirk makes it sound as if the Corbomite device is a small gadget hidden somewhere on the ship, in reality, half the spaceship would have to be made of Corbomite.[2]

At this point, someone might argue that the reaction of Corbomite does not necessarily have to occur in two steps. Corbomite could simply decompose into some other substances—without returning to its original state. Just as TNT, for example, decomposes into nitrogen, hydrogen, carbon monoxide, and carbon. In fact, that would be much more plausible. However, it would not fit Captain Kirk's claim. An explosive either explodes or it does not. A partial explosion does not work very well. This is exactly where the crucial difference lies.

A classic explosive could undoubtedly be ignited by enemy fire. To that extent, it would fit Captain Kirk's claim. However, once the decomposition of an explosive has begun, there is no stopping it. To trigger the explosion, energy must be supplied from the outside. This ignition spark sets off a chain reaction. Chemical reactions occur and release energy. A lot of energy! This energy ignites the remaining parts of the explosive. The strength of an explosion depends only on the amount of explosive. The size of the ignition spark does not matter. If the amount of energy supplied is too small, there is no ignition. Once the amount of energy supplied is large enough, ignition occurs. However, a larger ignition spark does not lead to a stronger explosion.[3]

If Corbomite is supposed to release as much energy as was used in the destruction of the spaceship, then it cannot simply be an exothermic reaction. Such a reaction would not simply stop when the amount of energy supplied is reached. Corbomite would therefore actually have to absorb the energy through an endothermic reaction first. This reaction would then have to run backward in a second step. In doing so, the previously absorbed amount of energy would be released again. Otherwise, the effect of Corbomite as outlined by Captain Kirk cannot be explained.

However, if Corbomite can decompose so easily and suddenly, then the question arises as to why it should immediately reform from its decomposition products afterward? It is quite possible for reactions to run forward and backward. This is actually more the rule than the exception in chemistry. The direction in which a reaction runs, however, depends on the prevailing conditions. In practice, these are usually the temperature and pressure. If conditions prevail that lead to the decomposition of Corbomite, then the reverse reaction should not occur immediately afterward. If the reverse reaction were to occur with a significant delay, then the threat of the Corbomite device would be rather weak. The attacker could move away in time.

---

[2]The amount of Corbomite could be significantly smaller if it were a nuclear reaction. Then, much more energy could be provided with significantly less material. Apart from the fact that Kirk's formulation sounds more like a classic chemical reaction, the considerations made would ultimately apply in the same way.

[3]Another question that arises is how Corbomite can distinguish between an attack and, for example, an ion storm? If the Enterprise really had a substance like Corbomite on board (or even consisted largely of it), it would constantly be in danger of accidentally destroying itself.

If Kirk had simply claimed that Corbomite would cause a massive explosion upon the destruction of the ship, that would have been much more plausible. However, this bluff did not seem threatening enough to him. Therefore, he apparently wanted to up the ante to really unsettle Balok. He was lucky that Balok apparently did not pay much attention in chemistry class. But as it turns out, Balok almost wanted to be deceived. As it turns out in the end, he is quite a friendly fellow and not really bent on destruction.

After this incident, James T. Kirk seems to have thought more thoroughly about the Corbomite matter. In doing so, he apparently noticed his mistakes. Nevertheless, the basic idea seems to have appealed to him, at least as a way out in otherwise hopeless situations.

Only one year later, he resorts to it again. In the 11th episode of the 2nd TOS season, *"The Deadly Years"*, he has only slightly revised his bluff. Due to a previously unknown disease, the crew of the Enterprise ages rapidly. As a result, Commodore Stocker takes command of the Enterprise. Although he may outrank Captain Kirk, he does not come close to the true captain in terms of competence. His inexperience in commanding starships eventually leads him to take the shortest route: through the Neutral Zone. This is known to be a bad idea. The Romulans are there in no time. The Romulans are, of course, not at all pleased. The outcome is predictable: they want to destroy the Enterprise.

Fortunately, Captain Kirk (now recovered) returns to the bridge just in time to save the day. And he resorts to his old trick. Only now it is a bit more sophisticated. He orders a message to be sent to Starfleet Command. Encrypted, but with a code that he knows the Romulans have long since cracked. In this message, he claims that he is forced to destroy the Enterprise and the Romulan attackers. And he will do so using the new Corbomite device. He still obviously has no idea what substance would cause such an explosion and what reaction would provide the corresponding energy. However, the Romulans apparently do not maintain diplomatic relations with the First Federation and thus hear about Corbomite for the first time. If they had just sat down and calculated how much energy a chemical substance can release at most, they could have realized that there is no danger to them at the distances between starships in space. But they are hearing about Corbomite for the first time and have to decide quickly. They prefer not to take the risk and withdraw.

Whether James Kirk (or another Starfleet captain) has resorted to the Corbomite trick more often (or will do so in the future) is not known. However, it is likely that it remained at these two instances. One should not rely too often on the opponent not paying attention in chemistry class.

6.2 The Molecule of Molecules

Even the Borg know something like longing. For their collective consciousness, there is one thing that represents absolute perfection to them. And perfection is everything to the Borg. That is why they not only conquer foreign civilizations.

They even force them into their collective in such a way that their minds become part of their shared consciousness. All just to learn everything from them. They have only one motivation: they want to become even better. They want to get even closer to absolute perfection. There is one thing that expresses all that they strive for, why they conquer and assimilate. The manifestation of absolute perfection at the highest complexity. This greatest thing that the Borg can imagine is nothing other than a molecule. The Omega molecule.

In the 21st episode of the 4th VOY season, *"The Omega Directive"*, Star Trek fans learn what even most Starfleet officers do not know. Over a hundred years ago, the Federation researched a molecule that was supposed to represent an inexhaustible energy source. There was only one problem: the molecule was unstable. It decayed and released an incredible amount of energy in the process. The amount of energy was not only so great that the involved scientists lost their lives. In an entire sector of space, the subspace was destroyed as a result. Even a century later, no starship can still fly there with warp drive. This dangerous knowledge was subjected to the strictest secrecy. Starfleet officers only learn about it upon their promotion to captain. The associated, top-secret regulation bears the name of this molecule: The Omega Directive. Named after the last letter of the Greek alphabet.

What kind of chemical is this Omega molecule? How can a single molecule contain as much energy as a warp core? And how can it destroy the subspace of an entire sector or even quadrant?

Let's start with the question of what the Omega molecule actually is. We don't learn much. However, Captain Janeway does show a picture of it once. Admittedly, a somewhat blurry picture. From the perspective of the episode's creators, this blurry depiction is quite understandable. After all, they themselves do not know exactly what an Omega molecule is supposed to look like. In this case, a not entirely sharp image is quite helpful. From a chemical perspective, blurry images of molecules are not only annoying. After all, we would like to know exactly what this super-molecule looks like. It is also very unusual for the image of a molecule to be blurry. Molecules are not photographed but drawn. We had already seen in one of the previous chapters why it is not so easy to take a photograph of a molecule where the individual atoms can be clearly recognized. For this reason, depictions of molecules are always drawings. Essentially, even very simple drawings. Molecules are represented as simple line drawings. The lines do not denote the atoms but the bonds between them. At the end of each line, unless otherwise specified, one must imagine a carbon atom. To maintain clarity, hydrogen atoms and their bonds are usually omitted. All other atoms are represented by the corresponding element symbol. For example, Cl for chlorine, N for nitrogen, or S for sulfur. The image from the Starfleet database that Kathryn Janeway shows her senior officers, on the other hand, is a somewhat blurry image of a spherical structure, in the center of which something else seems to be located.

Spherical molecules are indeed known. And they seem to have fascinated people even in more recent times. In 1996, there was the Nobel Prize in Chemistry for their discovery. Less than two years later, the mentioned Voyager episode *"The Omega Directive"* was filmed. The fact that there was even a Nobel Prize

6.2 The Molecule of Molecules

for the discovery of these spherical molecules suggests that they had a similar significance for earthly science in the late 20th century as the Omega molecule did for the Borg. However, they were not named Omega molecules but were called fullerenes.[4]

Fullerenes are a modification of carbon. Carbon is known to occur in various forms. This applies not only to the many chemical compounds it forms with other substances. Even elemental carbon exists in different forms, called modifications. The two most important and well-known are diamond and graphite. Both materials consist solely of carbon. One of them is crystal clear and incredibly hard. The other is black and quite soft. The difference lies solely in the chemical bonding. In diamond, each carbon atom is bonded to four other carbon atoms. The four bonds are oriented in such a way that they each assume the maximum angle between each other. This results in a structure known as a tetrahedron. One bond points upwards. One points forward and slightly downwards. One points, slightly backward, to the left and slightly downwards. And one points, slightly backward, to the right and slightly downwards. Depending on how you look at it, any of the four bonds can be any of the described bonds. The other bonds are then still arranged in the same way. This perfect structure with precisely oriented bonds makes diamond so hard.

In graphite, on the other hand, each carbon atom is bonded to only three other carbon atoms. Lying in a plane, this results in an angle of exactly 120 degrees between the bonds. Looking down on the plane, you have a pattern of interconnected, regular hexagons in front of you. Each of these planes is also very stable. But this only applies to the cohesion within a plane. There is no direct bond between the planes. However, this does not mean that the individual planes in graphite are not connected at all. Simply put, one of the four bonds of carbon is still left. Three of them, very similar to the four bonds in diamond, are to neighboring carbon atoms. The fourth, however, is not directly oriented to a single other atom. The electrons that should represent the fourth bond act between the individual planes. Because the electrons are not fixed in a so-called σ-bond between two atoms, they can move relatively freely. These electrons form a so-called π-system. This is the reason for the good electrical conductivity of graphite. These electrons also cause a certain cohesion between the individual layers of the carbon planes. Therefore, the individual planes of graphite do not simply fall apart.

If individual carbon layers are isolated from graphite, you get another carbon modification called graphene. Starting from graphene, further modifications can be derived conceptually. Let's imagine a layer of graphene and roll it up. We do not roll it up like a carpet. So, we do not form a spiral. Instead, we reconnect it to form a tube. This modification is called carbon nanotubes. Carbon nanotubes were discovered shortly after fullerenes and quickly attracted great interest. Among other

---

[4] The fullerenes are not named after one of their discoverers but after an architect: Richard Buckminster Fuller. In his architecture, geodesic domes played a significant role, which look a bit like giant fullerene molecules.

things, they were proposed as storage material for hydrogen. Hydrogen molecules were supposed to adsorb on the large surface of the carbon nanotubes. This process is called adsorption (with a "d", not with a "b"). After a promising initial publication on this at the end of the 1990s, various research groups took up the topic. They then published measurement results with even higher storage capacities. In the following years, however, the measured hydrogen storage capacities that were published became smaller and smaller. Around 2010, it was clear that the realistic storage capacities were very small. The high values from the older publications were apparently based on measurement errors. Carbon nanotubes are probably not suitable as hydrogen storage.[5]

If you do not simply roll up a graphene layer but instead curve it upwards in all directions, you get a sphere. The result is then a fullerene. Technically, fullerenes cannot be produced in this way. The carbon atoms in graphene are exclusively arranged in regular hexagons. These hexagons are planar. This means that only a plane can be formed with them. To achieve a curvature, pentagons are also needed. If you "unroll" a fullerene mentally, you get a network of carbon atoms consisting of regular pentagons and hexagons. Fullerenes come in different sizes. For example, they can consist of 60, 70, 76, 80, 82, or even up to 94 carbon atoms. One of the most well-known representatives is probably the fullerene made up of 60 carbon atoms. These are arranged in regular pentagons and hexagons in such a way that the molecule resembles a soccer ball.

A few more details

The Omega molecule in the depiction from the Starfleet database consists not only of a sphere. Additionally, there seems to be something in its center. This is reminiscent of a curious special form of fullerenes, known as an endohedral complex. Atoms of other elements can be enclosed inside a fullerene sphere. These are not bound to the inner surface of the sphere by classical chemical bonds. They are simply held in place because the foreign atom is trapped inside. In this way, a kind of "compound" of the noble gases helium or neon can be created, for example.

In this case, a helium or neon atom is trapped inside a fullerene. Since it is not a conventional chemical compound, a somewhat improvised notation is used. The corresponding endohedral complexes are written as $He@C_{60}$ or $Ne@C_{60}$. The endohedral complexes probably do not have any real practical use. However, this chemical gimmick does have a certain scientific coolness. ◄

We see that spherical molecules are indeed possible. So, the Omega molecule is not an unrealistic matter at first. The fascination of the Borg with it is also not entirely far-fetched. On 20th-century Earth, there was a similar fascination, as a

---

[5] See: Chang Liu, Yong Chen, Cheng-Zhang Wu, Shi-Tao Xu, Hui-Ming Cheng, "Hydrogen storage in carbon nanotubes revisited", *Carbon*, 2010, 48, 2, 452–455.

recent Nobel Prize shows. However, we do not know much more about the Omega molecule. We learn that a mineral called boronite is needed for its production. Boronite should not be confused with bornite. Bornite is a fairly common mineral on Earth from the sulfide group. This means that it is a salt whose anion is a doubly negatively charged sulfur atom. However, bornite does not only have a single cation but also contains a bit of iron in addition to copper. This bornite is obviously not related to boronite. The three elements contained in it are all nothing special. If iron, copper, or sulfur are present in the Omega molecule at all, they could just as well be obtained from many other sources.

Ultimately, we know even less about boronite than we do about the Omega molecule. We only know that it is needed as a raw material for Omega molecules. What does this tell us about boronite?

There could be two possible explanations for the role of boronite in Omega synthesis. Explanation one would be that the Omega molecule contains an element (still unknown to us today). Boronite would then simply be an ore from which this element is extracted. Explanation two would be that boronite contains precursors of the Omega synthesis. In principle, any chemical compound can be synthesized from the underlying elements. This synthesis would be very complex in most cases. Therefore, chemical syntheses often start with precursors. This means that compounds are used that can serve as building blocks for the larger and more complex target molecule. In nature, precursors for many chemical synthesis can be found. Some of these substances can be isolated from petroleum (which is another reason why it is not wise to burn large quantities of petroleum). Many complex intermediates can also be found in biology. If you do not have to start with the elements in a chemical synthesis but already possess essential structural elements of the target molecule, the synthesis is simply much easier. You save the first part of the process.

This would mean that boronite consists of molecules that are essentially fragments of an Omega molecule. You would save the effort of producing the fragment obtained from boronite. Instead, you could proceed directly to the actual synthesis of the Omega molecule. Unlike the case where boronite contains a necessary element, boronite would not be essential for the production of Omega molecules. You would just have to synthesize the corresponding intermediate yourself. However, it could very well be that the effort for the overall synthesis without this precursor would be so great that it becomes completely impractical.

So much for what we can say about the production of Omega molecules and their structure. However, the focus of the episode is not on their production but on their effect. Here, similar questions arise as with corbomite, which we discussed in the last section.

Omega molecules apparently contain an unimaginable amount of energy. No exact value is given for the energy content. We only learn that a single molecule contains as much energy as a warp core. Although we do not know how much energy is in such a warp core, it is obviously a lot. The decay of a single molecule seems to far overshadow the explosion of an atomic bomb. After all, the amount of energy can even destroy subspace. Can a single molecule contain so much energy? And if so, what would be the consequences?

Let's first look at the consequences. Here, what we learn in the episode is quite plausible. We hear that the molecule is very unstable and decays quickly. This fits with a very high-energy molecule. The stability of a chemical compound depends greatly on its energy level. The more energetic a substance is, the more unstable it is. Simply put, all substances strive for a minimum of energy.[6]

Let's imagine a simple example. Think of an oxyhydrogen mixture: a gas mixture of two parts hydrogen and one part oxygen. If we compare the energy content of the mixture with water, the product of their reaction, we find that the energy content of hydrogen and oxygen is significantly higher than that of water. This is noticeable by the large amount of heat released during the reaction. The oxyhydrogen mixture is quite chemically unstable. It can be ignited by a small spark. In doing so, it transforms into water and releases a lot of energy. Conversely, the product is very stable. Water does not simply decompose back into oxygen and hydrogen. While this is possible, a lot of energy must be supplied. The water molecule does not decompose on its own. Therefore, the reactants (hydrogen and oxygen) are significantly more unstable than the product (water). The difference is the different energy levels.[7] The same is true for the Omega molecule. This molecule contains a lot of energy and is correspondingly unstable. The Federation's and the Borg's problems in stabilizing the Omega molecules are therefore quite plausible.

It looks a bit different with the amount of energy. One wonders how a single molecule can contain so much energy. The most energy is usually released during total oxidation. This is nothing other than combustion. The amount of energy released during the combustion of a molecule depends on a number of factors. The essential point, however, is the number of oxidizable atoms in the molecule. For example, the combustion of an alkane with two carbon atoms (ethane, C_2H_6) releases an amount of energy of 1560 kilojoules per mole. If you burn an alkane with four carbon atoms (butane, C_4H_{10}), it is already 2877 kilojoules. With ten carbon atoms (decane, $C_{10}H_{22}$), almost 7000 kilojoules per mole are released. The series could be continued indefinitely. The more atoms an alkane molecule has, the more energy is released during its combustion. The Omega molecule is obviously

---

[6] The stable state at a certain temperature and pressure is reached when an energy quantity, known as Gibbs energy G (or free enthalpy), assumes the smallest possible value. This state is referred to as equilibrium. This is important, for example, when one wants to calculate the equilibrium constant. The equilibrium constant describes the composition a system has at reaction equilibrium. The equilibrium constant can be determined from the difference in Gibbs energies of products and reactants.

[7] As seen in the example of the oxyhydrogen mixture, a high energy level does not necessarily mean that a substance decomposes spontaneously. As in the case of oxyhydrogen, it first needs a trigger like the ignition spark. Such states are referred to as metastable. They inherently possess a very high energy level, and the transformation into the products would increase stability. However, the transformation chemically proceeds through intermediates that lie at an even higher energy level than the initial reactants. Thus, the high-energy oxyhydrogen is initially seemingly stable. However, if energy is supplied in a suited way, the intermediates can be formed in sufficient quantities. The reaction then becomes unstoppable and proceeds explosively.

6.2 The Molecule of Molecules

a very large molecule. You can't really see much on the display. However, it seems to be significantly larger than a fullerene. But even if it contained thousands or even tens of thousands of atoms, the amount of energy would hardly be enough to destroy subspace. Even if the Omega molecule contained about 60 trillion carbon atoms (that would be one mole and weigh 12 g), their combustion would release only about 393 kilojoules of energy. This would lift 1000 kg almost forty meters high on Earth. That is quite a lot. However, the energy content of a warp core is something else entirely.

The decay of the Omega molecule does not seem to be combustion. Most other chemical reactions, however, deliver less rather than more energy than combustion. The mere decay of an unstable molecule can release some energy. However, this amount of energy is manageable. Even if the molecule is very unstable, not an unlimited amount of energy is released. Chemical compounds simply do not contain enough energy for that. And what if one assumed that the decay of the Omega molecule is not a chemical reaction?

First of all, this explanation would not be really plausible. After all, the character as a molecule is repeatedly emphasized. Its chemistry must therefore play a central role. If it were a nuclear process, it would not be triggered by a chemical process (even though processes in atomic nuclei are dealt with in so-called nuclear chemistry). From boronite, some unknown element could theoretically be obtained, which is extremely energy-rich. Its decay or non-decay would, on the one hand, not be influenced by the molecule it is in. On the other hand, radioactive decays release a lot of energy. However, none of this would be comparable to the energy content of a warp core and could destroy subspace. Even if the special arrangement of atoms in the Omega molecule were such that the right elements were brought together so that the atomic nuclei would undergo a kind of nuclear fusion (however that would work): The amount of energy would still be manageable. Even with a very large molecule, relatively few nuclei would fuse. In today's hydrogen bombs, significantly more atomic nuclei fuse. And yet hydrogen bombs are not enough to destroy subspace.

The hypothetical upper limit for the amount of energy that could be released during the decay of Omega molecules is ultimately determined by Einstein's General Theory of Relativity. The equation $E = m \cdot c^2$ tells us, as is well known, that in the conversion of matter into energy, mass is decisive. Since the speed of light c is very high and its square even more so, a lot of energy is released even with very little mass. But even if we are dealing with a complete annihilation of matter, exactly as much energy is released as in the conversion of half the mass of antimatter with matter (half, because the mass of the antimatter is added to the mass of the matter). Even if Omega molecules may be quite large, the destruction of a solar system is of a different magnitude, and antimatter explosions are also known in Star Trek. They are dangerous, but not comparable to what we experience with the Omega molecule.

The last question related to the Omega molecule is similar to what we had with the Corbomite. What actually happens to the energy when the Omega molecules are destroyed?

The plan (ultimately carried out) is to destroy them with some devices from the 24th century and thus render them harmless. But the energy must go somewhere. The principle of energy conservation applies even in the Delta Quadrant. And for all molecules. It is conceivable that the Omega molecules could be decomposed in a controlled manner. This would slow down the decay accordingly. The energy would not be released abruptly. This would make sense, as it would prevent or at least limit the destruction that would otherwise occur. The energy would have to be dissipated under controlled conditions. However, the destruction of the Omega molecules on board the Voyager takes place within a few seconds. The speed of energy release cannot be particularly low. Where all this energy is supposed to flow in such a short time and how such power is to be controlled is another question that remains to be answered over the centuries until the time of Captain Janeway. Today's science and technology cannot yet provide an explanation for this.

Bibliography

Episodes of Star Trek, sorted by series (within the series sorted chronologically).
Citation style: Title; scriptwriter; director; season, episode number, first aired, (referencing chapter in this book).

TOS:

„*The Cage*"; Gene Roddenbery; Robert Butler; 1. pilot from the year 1964 (The Bussard Collector or Collecting from the Vacuum)
„*The Corbomite Maneuver*"; Jerry Sohl; Joseph Sargent; S 1, E 3, 1966 (Corbomite or Kirk's Favorite Chemical Bluff)
„*The Man Trap*"; George Clayton Johnson; Marc Daniels; S 1, E 5, 1966 (*The Salt Vampire of M-113; Why Does a Dead Shapeshifter Revert to Its Natural Form?*)
„*Arena*"; Gene L. Coon (Story: Fredric Brown); Joseph Pevney; S 1, E 19, 1967 (Corbomite or Kirk's Favorite Chemical Bluff)
„*The Devil in the Dark*"; Gene L. Coon; Joseph Pevney; S 1, E 26, 1967 (Horta or Life from Silicon)
„*Metamorphosis*"; Gene L. Coon; Ralph Senensky; S 2, E 2, 1967 (Life without a Body)
„*The Doomsday Machine*"; Norman Spinrad; Marc Daniels; S 2, E 6, 1967 (Materials That Do Not Consist of Chemical Elements)
„*The Deadly Years*"; David P. Harmon; Joseph Pevney; S 2, E 11, 1967 (Corbomite or Kirk's Favorite Chemical Bluff)
„*Obsession*"; Art Wallace; Ralph Senensky; S 2, E 18, 1967 (*Life without a Body; The Salt Vampire of M-113*)
„*By any other name*"; D.C. Fontana, Jerome Bixby (Story: Jerome Bixby); Marc Daniels; S 2, E 21, 1968 (A Thirsty Virus)
„*The Omega Glory*"; Gene Roddenberry; Vincent McEveety; S 2, E 25, 1968 (A Thirsty Virus)
„*Day of the Dove*"; Jerome Bixby; Marvin J. Chomsky; S 3, E 11, 1968 (Life without a Body)
„*The Savage Curtain*"; Gene Roddenberry, Arthur Heinemann (Story: Gene Roddenberry); Herschel Daugherty, S 3, E 22, 1969 (Very Hot Extraterrestrials)

TNG:

„Home Soil"; Robert Sabaroff (Story: Karl Guers, Ralph Sanchez, Robert Sabaroff); Corey Allen; S 1, E 18, 1988 (Horta or Life from Silicon)
„Coming of Age"; Sandy Fries; Mike Vejar; S 1, E 19, 1988 (The Mixing Ratio of Matter and Antimatter)
„Where Silence Has Lease"; Jack B. Sowards; Winrich Kolbe; S 2, E 2, 1988 (Life without a Body)
„The Dauphin"; Scott Rubenstein, Leonard Mlodinow; Rob Bowman; S 2, E 10, 1989 (Simply Being Someone Else)
„Night Terrors"; Pamela Douglas, Jeri Taylor (Story: Shari Goodhartz); Les Landau; S 4, E 17, 1991 (*One Moon Circles*)
„Ship in a Bottle"; René Echevarria; Alexander Singer; S 6, E 12, 1993 (Tiny Atoms—Part 2)

DS9:

„Emissary"; Rick Berman, Michael Piller (Story: Michael Piller); David Carson; S 1, E 1, 1993 (Simply Being Someone Else)
„Past Prologue"; Katharyn Powers; Winrich Kolbe; S 1, E 3, 1993 (Simply Being Someone Else)
„Melora"; Michael Piller, James Crocker, Steven Baum, Evan Carlos Somers (Story: Evan Carlos Somers); Winrich Kolbe; S 2, E 6, 1993 (Breathing Hydrogen)
„The Jem'Hadar"; Ira Steven Behr; Kim Friedman; S 2, E 26, 1994 (Materials That Do Not Consist of Chemical Elements)
„Homefront"; Ira Steven Behr, Robert Hewitt Wolfe; David Livingston; S 4, E 11, 1995 (Why Does a Dead Shapeshifter Revert to Its Natural Form?)
„The Muse"; René Echevarria (Story: René Echevarria, Majel Barrett-Roddenberry); David Livingston; S 4, E 21, 1996 (Life without a Body)
„Bar Association"; Barbara J. Lee, Jenifer A. Lee (Story: Ira Steven Behr, Robert Hewitt Wolfe); LeVar Burton; S 4, E 16, 1996 (Materials That Do Not Consist of Chemical Elements)
„Body Parts"; Hans Beimler (Story: Robert J. Bolivar, Louis P. DeSantis); Avery Brooks; S 4, E 25, 1996 (Materials That Do Not Consist of Chemical Elements)
„Ferengi Love Songs"; Ira Steven Behr, Hans Beimler; René Auberjonois; S 5, E 20, 1997 (Materials That Do Not Consist of Chemical Elements)
„Who Mourns for Morn?"; Mark Gehred-O'Connell; Victor Lobl; S 6, E 12, 1998 (Materials That Do Not Consist of Chemical Elements)
„One Little Ship"; Bradley Thompson, David Weddle; Allan Kroeker; S 6, E 14, 1998 (Tiny Atoms)
„What You Leave Behind"; Ira Steven Behr, Hans Beimler; Allan Kroeker; S 7, E 27, 1999 (Materials That Do Not Consist of Chemical Elements)

VOY:

„The Cloud"; Michael Piller, Tom Szollosi (Story: Brannon Braga); David Livingston; S 1, E 6, 1995 (Life without a Body)
„Emanations"; Brannon Braga; David Livingston; S 1, E 9, 1995 (How many Elements are there Actually?)
„Learning Curve"; Ronald Wilkerson, Jean Louise Matthias; David Livingston; S 1, E 16, 1995 (Materials That Do Not Consist of Chemical Elements)

"*Threshold*"; Brannon Braga (Story: Michael DeLuca); Alexander Singer; S 2, E 15, 1997 (Crossing the Threshold)

"The Gift"; Joe Menosky (Story: Kenneth Biller, Jack Klein, Karen Klein, Scott Nimerfro, James Thornton); Anson Williams; S 4, E 2, 1997 (When Atoms Burn)

"Scientific Method"; Lisa Klink (Story: Sherry Klein, Harry Doc Kloor); David Livingston; S 4, E 7, 1997 (*Tiny Atoms—Part 2*)

"The Omega Directive"; Lisa Klink (Story: Jimmy Diggs, Steve J. Kay); Victor Lobl; S 4, E 21, 1998 (*The Molecule of Molecules*)

"Demon"; Kenneth Biller (Story: André Bormanis); Anson Williams; S 4, E 24, 1998 (*Simply Being Someone Else*)

"*Bride of Chaotica!*"; Bryan Fuller, Michael Taylor (Story: Bryan Fuller); Allan Kroeker; S 5, E 12, 1999 (Life without a Body)

ENT:

"*Rogue Planet*"; Rick Berman, Chris Black, Brannon Braga (Story: Chris Black); Allan Kroeker; S 1, E 18, 2002 (Simply Being Someone Else)

"*Regeneration*"; Phyllis Strong, Mike Sussman; David Livingston; S 2, E 23, 2003 (The Bussard Collector or Collecting from the Vacuum)

"Observer Effect"; Judith Reeves-Stevens, Garfield Reeves-Stevens; Mike Vejar; S 4, E 11, 2005 (Life without a Body)

DSC:

"Choose your pain"; Kemp Powers (Story: Gretchen J. Berg, Aaron Harberts, Kemp Powers); Lee Rose; S 1, E 5, 2017 (A Thirsty Virus *and* Simply Being Someone Else)

"Saints of Imperfection"; Kirsten Beyer; David Barrett; S 2, E 5, 2019 (Explosions in Space)

PIC:

"*Maps and Legends*"; Michael Chabon, Akiva Goldsman; Hanelle M. Culpepper; S 1, E 2, 2020 (Materials That Do Not Consist of Chemical Elements)

Movies:

"*Star Trek VI: The Undiscovered Country*"; Nicholas Meyer, Denny Martin Flinn (Story: Leonard Nimoy, Lawrence Konner, Mark Rosenthal); Nicholas Meyer; 1991 (The Bussard Collector or Collecting from the Vacuum *and* Simply Being Someone Else)

"*Star Trek: Insurrection*"; Michael Piller (Story: Rick Berman, Michael Piller); Jonathan Frakes; 1998 (Explosions in Space)

The manufacturer's authorised representative in the EU is Springer Nature Customer Service Centre GmbH, Europaplatz 3, 69115 Heidelberg, Germany. If you have any concerns regarding our products, please contact ProductSafety@springernature.com

Printed and bound by CPI Group (UK) Ltd, Croydon, CR0 4YY

26/03/2026

02078952-0019